CW01510050

Phlomis

The Neglected Genus

A Guide for Gardeners and Horticulturists

Jim Mann Taylor

J Mann Taylor ◆ Gloucester

First published July 1998
Copyright © general text Jim Mann Taylor 1998
Copyright © photographs Jim Mann Taylor & Jean-Pierre Jolivot 1998

All rights reserved. No part of this
publication may be reproduced,
in any form or by any means,
electronic or mechanical,
without permission of the author.

Published by:
J. Mann Taylor
Sunningdale
Grange Court
Westbury-on-Severn
Glos. GL14 1PL
in association with the NCCPG

A catalogue record of this book
is available from the British Library

ISBN 0-9532413-0-0

Contents

Acknowledgement

My grateful thanks to Ian Hedge of the Royal Botanic Garden, Edinburgh for patiently answering numerous questions over the years.

Key:

✦ means that the plant is in the NCCPG National Collection.

✧ means that the plant is in the French National Collection.

❀ means a good garden plant in the author's opinion.

❀❀ means an outstanding garden plant in the author's opinion.

❄ means that the plant may be killed with sub-zero (°C) temperatures.

Introduction

The NCCPG National Collection of *Phlomis* differs from many other collections in that it is mostly composed of species rather than cultivars. This has an important effect on the descriptions given. Cultivars are reproduced vegetatively by cuttings etc., and therefore all have similar characteristics. Species by their nature vary considerably. Consider mankind. We are all one species and yet some of us are tall, some short; some fat, some thin; some die at birth others live to 100; some go through life without a disease, others are crippled by diseases, many of them inherited.

So it is with plant species in the wild. Within a wild population of a particular species there will usually be considerable natural variation. If it is a fairly rare species its description in a Flora will depend very much on the field notes taken at the time by the collector. Back at the herbarium, the taxonomist may have one or several dried specimens to work on. Without good field notes the description at worst, may be that of just one plant. With a common species of course, there will be many collections from different populations in different areas and countries. The taxonomist then has a much better idea of natural variation and can build this into his Flora description. The more time one spends in the field the more one is amazed by the natural variation found within any one species.

One of the more surprising things about species in cultivation is that in many cases they may be descended from one original wild seed collection and have since been propagated vegetatively giving the appearance of uniformity. A species collection therefore could be the continued propagation of a single introduced species, or a variety of wild collected seed may be grown and a small measure of the natural variation maintained within cultivation. The *Phlomis* collection follows the latter.

The word *Phlomis* in history

Phlomis is an ancient name. Dioscorides (the Greek physician of the first century) distinguished two groups of plants 'phlomos' and 'phlomis', each further subdivided.

'Phlomos'—*The principal division of this is twofold; there is a white kind and a black kind, and of the white kind there is a female and a male. The leaves of the female kind closely resemble the cabbage, but are much hairier, flatter, and whitish; the stem is more than a cubit tall, whitish, somewhat hairy; the flowers are white tinged yellow, the seed is black; the root is long, rough, a finger's breadth across; it grows in the plains. The other kind, that called male, is whitish-leafed, somewhat elongated, with narrower leaves and a more slender stem. There is also a wild phlomos with tall shrubby shoots and leaves like wild sage; it bears on its shoots whorls of quince-yellow flowers resembling those of white horehound.*

There are also two kinds of 'phlomis' which are hairy, prostrate, and with curved (or perhaps *'compact'*) *leaves; a third 'phlomis', that called lychnitis (by some called thryallis), has three or four or more small leaves which are thick, hairy, greasy, and useful for making lamp-wicks.* (personal communication from R.C. Palmer)

In the above descriptions, the wild tall shrubby 'phlomos' with yellow whorls like horehound sounds not unlike *P. fruticosa* L., (sometimes referred to as Jerusalem sage, although it does not grow in Palestine). *Phlomis lychnitis* being used for lamp wick sounds valid for *P. lychnitis* L. as it was used for this purpose in Spain where it was known as Candelera (Curtis, 1807). However, the description of the leaves as three or four does not fit *P. lychnitis* L., and others have suggested it is the description of a *Verbascum*.

In a parallel passage in Pliny, the Greek 'phlomos' is largely equivalent to the Latin *Verbascum* or mullein, the white 'phlomos' denoting white-woolly species such as *Verbascum thapsus*, the black 'phlomos' being a less woolly species such as *V. nigrum* and *V. lychnitis*.

The first description of the *Phlomis* genus as we know it today, was in Tournefort's *Compleat Herbal* of 1716; the first valid publication was by Linnaeus in his *Species Plantarum* of 1753. He later recognised twelve species, seven of which are still recognised as *Phlomis* today. Miller (1768) in his *Gardener's Dictionary* described 14 species, at least six of which have since been transferred to other genera.

General description of the genus *Phlomis*

Phlomis are herbaceous plants, sub-shrubs or shrubs, hardy or tender in the UK. Leaves are entire, opposite (each leaf pair at right angles to the next) and rugose or reticulate veined. The floral leaves or bracts are similar or different to the lower leaves. All parts are frequently covered with hairs, which are mainly stellate or dendroid and which can take a variety of forms including glandular. These are illustrated in Appendix 5. The flowers are arranged in whorls (verticillasters) round the stems which are usually square in section with rounded corners, although indumentum on the stems can make them appear more circular. The colour of the flowers varies from yellow to pink, purple and white. The bracteoles which if present, directly surround the whorl of calyces, are ovate, lanceolate or linear. The calyx is tubular, barrel- or campanulate-shaped with 5 or 10 veins visible. The 5 calyx teeth are either all of equal length or the outer two are longer than the others. The teeth are held at various angles. There are 4 stamens ascending under the upper lip. The anther has a forked end, the upper fork being shorter than the lower. The fruits are four three sided nutlets, sometimes topped with minimal hair, sometimes hairless. The root system can be very extensive, occasionally with tubers.

Notes on Classification

Phlomis belongs to the family *Labiatae*, subfamily *Lamioideae*.

The problem with the classification of *Phlomis* has been that taxonomists have not always used the same taxa when making their classifications. The most recent classification by Kamelin & Makhmedov (1990) is the most universal and has combined most of the Mediterranean taxa with those of Iran, Central Asia and China, but he goes down to sub-sections of one or two taxa which is not particularly helpful for horticulturists.

Moench (1794) recognised two separate genera *Phlomis* and *Phlomoides*. Link (1829) recognised two genera *Phlomis* and *Phlomidopsis.* Bentham (1834) divided the genus *Phlomis* into two sections *Euphlomis* and *Phlomidopsis* based on the corolla hood shape and the presence or not of a bearded fringe to the upper lip. Boissier (1879) adopted much of Bentham's classification, but with different sub-sections. Briquet (1895-97) returned to the name *Phlomoides* instead of *Phlomidopsis.* Vierhapper (1915) & Post (1933) generally followed Briquet, The above classifications worked mainly with European and middle-eastern species. Komarov (1954) used mainly Russian species. Later classifications by Kamelin & Makhmedov (1990) embraced USSR, Chinese, middle-eastern and European species. In this revision and the earlier one by Adylou, Kamelin & Makhmedov (1987), many of the species that previously were in Section *Phlomoides* were raised to the level of genus and were called *Phlomoides.* The genus *Eremostachys* was also included in *Phlomoides.* [*Eremostachys* is very closely related to *Phlomis* and the boundary between the two genera is not always clear; most *Eremostachys* have a well-developed tuft of hairs at the apex of the nutlets; the genus contains about 40 species concentrated in the Central Asiatic/ Afghanistan area. Few *Eremostachys*, many of which are very handsome plants, are in cultivation in the UK.] *Phlomis tuberosa* (the only Mediterranean species affected) was given the name *Phlomoides tuberosa*, a name it had been given nearly 200 years before by Moench.

This split into *Phlomis* and *Phlomoides* either as sections of the genus *Phlomis* or as separate genera, is underlined by the fact that chromosome determinations in the literature give in general 2n=20 for section *Phlomis* and 2n=22 for section *Phlomoides*.

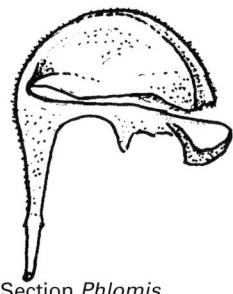

Section *Phlomis*   Section *Phlomoides*

Section *Phlomis*

Shrubs and herbaceous perennials. Upper lip of corolla large, hooded, somewhat hemispherical, strongly compressed laterally, to form a narrow ridge at the top, not densely bearded at inside margin (hairs sparse and short); lateral lobes of lower lip small, the middle lobe large and broadly rounded.

Section *Phlomoides*

Herbaceous perennials. Upper lip of corolla smaller, gently rounded, not strongly compressed laterally, and so not exhibiting the narrow ridge at the top, densely bearded at margin; lateral lobes of lower lip scarcely smaller than the middle lobe, obtuse.

The section *Phlomis* has often been split into sub-sections the most useful of which are Bentham's three sub-sections, *Gymnophlomis*, *Dendrophlomis*, and *Oxyphlomis*.

General Distribution

Phlomis (including sections *Phlomis* and *Phlomoides*) are found in most countries bordering the Mediterranean, both north and south (37 *Phlomis* + 1 *Phlomoides*) plus Turkey (27 *Phlomis* + 1 *Phlomoides*), Iran/Iraq/Afghanistan/Himalayas (27 *Phlomis* + 3 *Phlomoides*), former USSR (27 *Phlomis* + 22 *Phlomoides*) and China (1 *Phlomis* + 42 *Phlomoides*), plus numerous natural hybrids.

The members of *Phlomis* sub-section *Dendrophlomis* are shrubby and densely hairy with small leaves and corrugated leaf laminae. They are associated with the sclerophyllous vegetation of the maquis and garigue along the Mediterranean coast and grow in stony or rocky ground. *P. fruticosa* on the other hand has a much wider distribution into the former USSR and China. Many members of *Phlomis* sub-section *Gymnophlomis* and *Oxyphlomis* occur in Turkey and Iran. The region is characterised by a mild wet winter and a hot dry summer.

The members of section *Phlomoides* are to be found in the Irano-Turanian region, the majority in Soviet Central Asia and China. They grow in grassland and grass-steppe between the arid desert areas and also in the mountain ranges of the Pamirs and the Tien-Shan. The region is characterised by low rainfall, hot dry summers, cold winters and melting snows in the spring which start the plants into growth. The plants of this group are generally herbaceous with membranous leaves and are generally less hairy and more mesophytic than those of section *Phlomis*. There is one member of this section present in the Mediterranean area, namely *Phlomis tuberosa* L. [syn. *Phlomoides tuberosa* (L.) Moench]. This grows in grassland and rocky steppe at 1000–1500m in countries such as Hungary, the former Jugoslavia, Bulgaria and Greece..

Naming

As with most genera, it comes as no surprise to find much confusion over naming. This is not confined to the UK or the nursery trade. For example, the results of a comparison of *Phlomis* species offered to Marburg Botanic Garden between 1963 and 1965 (Ludwig, 1968), showed many mis-identifications. Six gardens were growing representatives of other genera of *Labiatae* as *Phlomis* species, and one even sent *Collomia grandiflora*, a member of the *Polemoniaceae*.

P. russeliana was received from 24 botanic gardens under the names of 6 other species:

> 10 under the name *P. samia*
>
> 9 under the name *P. viscosa*
>
> 2 under the name *P. alpina*
>
> 1 each under the names *P. pungens, P. capitata* and *P. ferruginea*
>
> Those from only 4 botanic gardens (14%) were received correctly named.

P. tuberosa was received from 32 botanic gardens under the names of 9 other species:

> 17 under the name *P. alpina*
>
> 4 under the name *P. herba-venti*
>
> 3 under the name *P. chrysophylla*
>
> 3 under the name *P. samia*
>
> 1 each under the names *P. pungens, P. cashmeriana,*
>
> *P. americana, P. armeniaca* and *P. lanata.*
>
> Those from 76 gardens (70%) were received correctly named .

In the UK a somewhat similar pattern occurs with the plants offered in the retail trade. The greatest confusion occurs with *P. samia* and *P. viscosa* for which the trade often supplies *P. russeliana*; *P. russeliana* was recently (July 1997) called *P. samia* on BBC's Gardeners World. The true *P. viscosa* and *P. samia* are of course still fairly rare in the nursery trade here. *P. lanata* and *P. leucophracta* seem to be reasonably free of errors (other than spelling errors, *P. leucophracta* as *P. leucopacta* for example.)

Why the great confusion? Well part of the reason is undoubtedly that the genus is a difficult one botanically, but synonymy plays a significant part. Boissier considered *P. russeliana* as a synonym of *P. viscosa* Poir., while at the same time confusing the true *P. russeliana* in his herbarium with *P. samia* L. (Huber-Morath 1958). The name *russeliana* is after Dr. Russell who illustrated a plant in his Natural History of Aleppo (Dr. Russell had considered it might be a yellow-flowered version of *P. herba-venti*). Curtis's Botanic Magazine (1825, tab. 2542) describes what we now know as *P. russeliana* under the name *P. lunarifolia* Smith var. *russeliana* Sims, "Russell's Honesty-leaved *Phlomis*". Bentham in 1834 revised the taxonomy and upgraded its status to species *P. russeliana* (Sims) Benth. This is where some of the confusion arises as nurseries rarely give the author citation and even some Botanic Gardens don't use them on their labels. *P. viscosa* Poir. (a yellow flowered glandular shrub) and *P. samia* L. (glandular, herbaceous perennial with purple flowers) are distinct plants in their own right, compared to the yellow flowered, non-glandular and perennial *P. russeliana* which is found only in northern Turkey.

Phlomis lunariifolia Smith was described (as *P. lunarifolia* Sibth. & Smith) in the Botanical Magazine (1900, tab. 7699) In 1905 H.S.Thompson, after borrowing Sibthorp's type specimen of *P. lunariifolia* from Oxford Botanic Herbarium, decided that the plant shown in tab.7699 was a species distinct from *P. lunariifolia* Smith. He gave it the name *P. grandiflora* H.S.Thomps.

Phlomis italica has an unfortunate name as the plant is endemic only to the Balearic Islands. It has been attributed to Morocco in one sheet of Montpelliers herbarium, but the reference is considered incorrect. (Mateu, 1986). Linnaeus who first applied the epithet *italica* may have received specimens or seed of the plant via Italy. An attempt was made in 1905, which saw the Second International Botanical Congress in Vienna, to change the specific epithet to *balearica*, but the Congress insisted on the use of the first correctly published name, a rule which has annoyed some gardeners ever since.

The plant *Phlomis bovei* subsp. *maroccana* is currently causing confusion in the UK Nursery trade. *Phlomis bovei* De Noé was described from North Africa. *Phlomis samia* L. was described from the eastern Mediterranean. Maire who first described *P. bovei* subsp. *maroccana* in 1928, changed his mind in 1934 including the three taxa, at subspecific rank under *P. samia*; hence the use of the name P. *samia* subsp. *maroccana* for the sub-species. Other botanists have not accepted this, insisting that *P. samia* L. and *P. bovei* De Noé are two different species so *P. bovei* subsp. *maroccana* it is for the present. (see my own views under the species write-up, p. 39).

Phlomis fruticosa has the common name Jerusalem Sage, which is unfortunate as one of the places it does not grow is in Israel or Jordan.

Phlomis lychnitis has the common name Lamp Wick (Spanish name Candelera). The Botanical Magazine (1807, tab. 999) says that the slender radical leaves are used as a wicks for lamps, 'for which purpose they are said to answer very well'. More recently however, Compton (1987) relates that *Phlomis* plants in general got their name from the use of the calyces, which are like small felted cones, as candles; the calyces were soaked in dishes of olive oil, and then lit at the tip. I have tried treating the larger *Phlomis* whorls in this way and it certainly works.

Pollination

The corolla of section *Phlomis* species consists of a three lipped lower segment and a hooded helmet-shaped upper segment. Both of these, because of their construction, are very difficult to deform. To gain access to the corolla, it is necessary to force the upper hinged lip upwards. In the case of many *Phlomis*, the only insect capable of this is the bumblebee, which starts by edging in from the side near the hinge, and then aligning itself with the main axis. (Brantjes, 1981).

P. bovei De Noé subsp. *maroccana* Maire generally produces more viable seed than other species in the UK. It has a large gap (around 5 mm) between the upper and lower lip, allowing easy frontal access by bees.

Phlomis species hybridise readily in the wild. The hybrids of *Phlomis*, as far as is known, are fertile, and so it is much more difficult in the field and especially with herbarium samples, to recognise them. *P. kurdica* × *P. oppositiflora* however, is easily recognisable, in spite of the intergrading series in an abundant population, because *P. oppositiflora* only ever has two flowers in each whorl, whereas *P. kurdica* has a considerable number. (Huber-Morath, 1958).

Propagation

Because they hybridise, the preferred method for commerce should be by cuttings or division. For the shrubby species, take half-ripe cuttings in June, July or August. For the herbaceous species, take divisions in spring or autumn or soft basal cuttings in late spring. Although many of the plants look as if they might rot off under mist due to their grey woolly surfaces, most root perfectly well under mist (if reduced to a minimum) in spring and summer, and don't need any rooting hormone. Most *Phlomis* are easily propagated from wild collected seed which germinates within two weeks. Seed can still be viable after some years (kept in the fridge), but I am becoming convinced that most field collectors gather *Phlomis* seed too early in the summer and that if the collection of the seed is delayed till around October, far better germination is achieved. Seed collected this way in the wild has germinated like mustard and cress even after two years.

Health and Safety

With *Phlomis*, many of which are covered in small, complex loosely held hairs, I find that my throat is affected when they are handled for a long time in dry conditions under cover. The feeling is rather akin to working in a house-loft lined with fibreglass. The effect fortunately quickly disappears and I have noticed no long term effects. A friend however, who suffers from asthma, when digging up a sample of *P. samia* for me in a poly tunnel, was rendered almost unconscious by the inhalation of hairs. Those handling the material on a long term basis under cover, such as nurserymen, would therefore be well advised to wear a face mask.

Cultivation

The majority of the species in cultivation in the UK are from the Mediterranean area and therefore are happy in full sun and good drainage. Although most of the shrubs are happy in droughts when established, young plants need watering until a suitable root system has been established. Many of the herbaceous plants on the other hand, particularly those from Central Asia and the Himalayas, need moisture in the spring (the equivalent of melting snows) to start them off and are best suited to a well cultivated border. *Phlomis* should be given an open sunny position here, although in nature many of the plants grow in the semi-shade of *Pinus halepensis* (syn: *brutia*) for example in Turkey, and in Greece with *Quercus coccifera*. Heavy shade in this country will cause substantial changes to the shape of the leaves and may lead to arguments as to which species it is. *P. longifolia* for example can be unrecognisable in heavy shade. Equally those species with a dense indumentum on the leaves often develop winter and summer leaves. In winter they may develop larger leaves with less indumentum compared with the smaller thickly felted summer leaves. This happens with *P. nissolii* and is illustrated by *P. brachyodon* in the following illustrations from Desert Vegetation of Israel & Sinai by Avinoam Danin (Cana Publishing House, 1983)

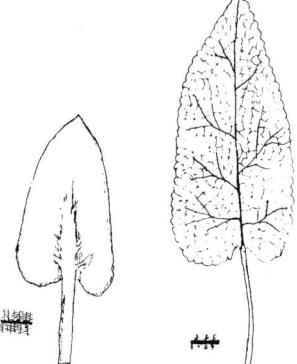

Winter leaves on right. Cross section of lamina shown for each leaf.

Philip Miller grew many different *Phlomis* in the Chelsea Physic Garden and records only two (*P. samia* and *P. orientalis*) that were wiped out (throughout England) in the winter of 1740. The Met. Office tell me that they have calculated monthly temperatures by various means from 1659 for Central England. The winters have been seasonally ranked and the coldest winters were 1683/84, 1739/40 and 1962/3. [1739/40 being colder than 1962/63]. So the fact that in January 1740 mention is made of only two *Phlomis* being exterminated, lends weight to the suggestion that they are hardier than many people believe.

Philip Miller says that his *Phlomis* plants rarely lived more than 12 to 14 years. On the other hand, Christopher Lloyd in his garden at Dixter has a *P. fruticosa* (now massive) which is some 45 years old.

The pH of the soil is not critical except for *P. atropurpurea* which prefers acid soil and very wet conditions. *P. angustissima, P. bourgaei* and *P. lycia* grow naturally on both acid and alkaline soils, and most species seem happy on both types of soil in cultivation.

Pests & Diseases

The main pest that seems to attack *Phlomis* is the Leaf Hopper (in the open and in the greenhouse). The leaves become speckled or drained of green. This can be controlled with applications of Heptenophos/ Permethrin. I have not found any *Phlomis* which is allergic to this mixture.

Slugs and snails will eat the leaves particularly of the perennial species like *P. russeliana* and *P. tuberosa*.

Other than these, the plants seem to be pest and disease resistant.

Medicinal or herbal uses

Philip Miller reported that the leaves of *P. fruticosa* and *P. angustifolia* have been greatly recommended by some persons to be used as 'tea for sore throats'. In Spain the leaves of *P. x composita* are being used externally for wound healing; the flowers and leaves of *P. lychnitis* internally for antidiarrhoeal, prostate treatment, fever reducing, and as a tranquilliser; the aerial part with flowers of *P. purpurea* internally for liver and kidney complaints, prostate treatment, tranquilliser and externally for wound healing and an antiseptic. Young shoots of *P. umbrosa* are used in Korea as a vegetable. The roots of *P. umbrosa* and *P. koraiensis* are used for 'black and blue' skin wounds. Those of *P. betonicoides* are used for colds and diarrhoea. Those of *P. maximowizii* are used to reduce swellings. The tubers of *P. tuberosa* are eaten in the former USSR.

Illustrations

Most illustrations of *Phlomis* are to be found in the *Botanical Magazine*, the *Botanical Register* or Sweet's *British Flower Garden*. Many of these are good illustrations and descriptions. These include:

Phlomis purpurea L.   *Bot. Mag.* t.518 (1968)

Phlomis italica L.   *Bot. Mag.* t.9270 (1929)

Phlomis cashmeriana Royle ex Benth.   *Bot. Reg.* t.22 (1844)

Phlomis spectabilis Falconer ex Benth.   *Bot. Mag.* t.8870 (1920)

Phlomis herba-venti L.   *Bot. Mag.* t. 2449 (1824)

Phlomis herba-venti L.   Sweet's *British Flower Garden* t.74 (1831-8)

Phlomis pungens Willd.   Sweet's *British Flower Garden* t.33 (1823-25)

Phlomis fruticosa L.   *Bot. Mag.* t.1843 (1816)

Phlomis tuberosa L.   *Bot. Mag.* t.1555 (1813)

Phlomis lychnitis L.   *Bot. Mag.* t.999 (1807)

Phlomis armeniaca Willd. Sweet's *British Flower Garden* t.364 (1831-8)

Phlomis floccosa D.Don *Bot. Reg.* t.1300 (1829)

Phlomis rigida Labill. is beautifully illustrated in Labillardière's *Icones Plantarum Syriae rariorum* 1791

'Errors' in Illustrations

Several *Phlomis* are now considered wrongly illustrated in the literature, and the following indicates the current correct names for those illustrations.

Phlomis grandiflora H.S.Thomps. is illustrated under the name *Phlomis lunarifolia* Sibth. & Smith in *Bot. Mag.* t.7699 (1900)

Phlomis bovei De Noé is illustrated under the name *Phlomis samia* in *Bot. Mag.* t.1891 (1818)

Phlomis russeliana (Sims) Benth. is illustrated under the name *Phlomis lunarifolia* var. *russeliana* in *Bot. Mag.* t.2542 (1825)

In the *Bot. Mag.* t.9144 (1926) *Phlomis bovei* De Noé is illustrated by what has since been named as a subspecies, *P. bovei* De Noé subsp. *maroccana* Maire. It is an excellent description of the sub-species.

A guide to identifying *Phlomis* using this book.

It is considered that the sub-section divisions of Bentham are not too useful to the gardener or nurseryman, so in this book the genus has been separated into two main and easily separated sections *Phlomis* and *Phlomoides*. Section *Phlomis* has been further divided into two parts, part 1, containing all the shrubs and part 2, the perennials. Each section is arranged alphabetically. Appendix 1 can be used as an Index to the pages of this book.

Included in the book are all the species known to be in cultivation either in gardens or in Botanic Gardens. [If anyone finds omissions the author would be grateful to be informed.]

Shrub or Perennial?

The first thing to do when faced with a *Phlomis* identification then, is to decide whether the *Phlomis* in question is a shrub or a perennial.

If it is a shrub they will be found in Section *Phlomis* Part 1—Shrubs

If your *Phlomis* is a perennial, you first need to determine whether it has the large hemispherical, ridged, upper lip to the corolla with a large central lobe to the lower lip of Section *Phlomis*; or the Section *Phlomoides* smaller more rounded upper lip, heavily bearded, with the lobes of the lower lip more equal. The perennials of the section *Phlomis* will be found in *Phlomis* Part 2—Perennials. The perennials of the Section *Phlomoides* will be found under the heading Section *Phlomoides*.

If you know the plant has been wild collected then consult the country listings at the end of the book. This will usually greatly narrow the choice.

The next thing to study is the flower colour. Amongst the shrubs, all have yellow flowers except *P. italica, P. purpurea* and *P. elliptica*. The upper lip of the flowers of *P. leucophracta* usually have a rust brown or reddish colour.

Amongst the perennials (*Phlomis* and *Phlomoides* sections), most are pink/purple/white. The following are yellow flowered: *P. angustissima, P. armeniaca, P. aucheri, P. brachyodon, P. bruguieri, P. bucharica , P. cancellata, P. capitata, P. crinita , P. kurdica, P. linearis, P. lychnitis , P. nissolii , P. oppositiflora, P. platystegia, P. russeliana* and *P. syriaca*. The flowers of *P. crinita* and its hybrids often have a distinctly brown upper lip to the corolla.

Species Descriptions

In the following lists, the author and botanical reference is given for each species, sub-species, etc. [Author names are given in full in the Index beginning on page 70.] Its distribution in the wild is then outlined. Flowering time in the wild is detailed if known. The meaning of the specific epithet is explained. Comments about its look and habit follow with an indication of how it differs from certain similar looking species. Special comments about propagation may be included. Drawings of the leaf, calyx and bracteoles together with any other diagnostic features are included where possible. Indented is a botanical description together with details of any natural hybrids.

In the botanical descriptions (the example given is for perennial plants) I start with the basal leaves; then the stem leaves of the flowering stem; whether the stem is simple and unbranched or heavily branched; the floral leaves (or bracts) at the base of each whorl; the number of whorls per stem and whether they are close together or distant; the diameter of the whorl complete with corollas if known; a description of the bracteoles; the calyx including teeth; the corolla colour and whether the nutlets have hairs at their apex.

For this genus, descriptions such as tomentose, pannose or lanate indicate that the surface being described has a dense covering (usually of stellate or dendroid hairs) such that to the naked eye the surface appears grey, white, yellow or silver, felted and woolly. If described as floccose, the outer layer will be easily removed in tufts by a finger. Viscid indicates that the surface will be sticky caused by the exudation from glandular hairs.

Note the difference between for example *linear-lanceolate* (which means the leaves are of unvarying shape somewhere between linear and lanceolate) and *linear to lanceolate* (which means some leaves are linear and others lanceolate). A glossary of the botanical terms used for describing *Phlomis* is included at the end of the book (Appendix 3).

Symbols and Descriptions

✦   means that the plant is in the NCCPG National Collection.

✧   means that the plant is in the French National Collection.

❀   means a good garden plant in the author's opinion.

❀ ❀   means an outstanding garden plant in the author's opinion.

❄   means that the plant may be killed with sub-zero (°C) temperatures.

Scale Note: None of the drawings in the text are to scale (except for the relative proportions of the calyx to the bracteoles). Reference should be made to the size descriptions within the text.

Section *Phlomis* Part 1—Shrubs

This listing contains descriptions of all Phlomis shrubs — alphabetically.

Phlomis amanica Vierh.

Öst. Bot. Zeitschr. 65:213, t. 6 f.2 (1915)
Synonym:
    P. chrysophylla Boiss. var. *oblongifolia* Boiss.
Distribution in the wild:
    Endemic to Turkey, on slopes around 90m.
Flowering in the wild: July
The specific epithet *amanica* means of the Amanus mountains, S. Turkey.
This shrub is only known from the original collection yielding the description below, and is therefore **extremely rare** and may never be collected again.
Similar to *P. monocephala* in its habit and leaf shape and hairiness. The hairs of the calyces and bracteoles are only stellate in *P. amanica*, but in *P. monocephala* are long, undivided and form a very dense covering.

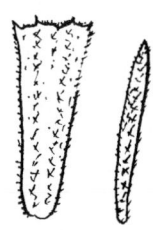

Non-glandular evergreen shrub to 100 cm. Leaves with short stellate indumentum, loosely stellate-hairy above, densely yellowish or whitish stellate-tomentose below. Cauline leaves oblong-ovate, obtuse, cuneate-truncate at base, crenulate at margin, 2.5–5 × 0.7–3 cm; petiole to 2 cm. Floral leaves oblong, shortly petiolate. One to two whorls, 6–12 flowered. Bracteoles linear, subulate, 10–15 × 1–2 mm, stellate-tomentose, without undivided hairs. Calyx tubular; teeth triangular-subulate, to 1 mm. Corolla yellow, 30 mm. Nutlets hairless.

Phlomis aurea Decne.

Ann. Sc. Nat. Bot. Sér. 2,2:251 (1834)
Synonyms:
    P. angustifolia Mill.
    P. flavescens Mill.
Distribution in the wild:
    Sinai, amongst rocks. Possibly in Jordan.
Flowering in the wild: March–June
aurea means golden and refers to the indumentum, especially on the young leaves.
Very rare. Said to roughly resemble *P. fruticosa,* but with a golden yellow indumentum in many parts, calyx teeth are very short and erect, middle lobe of lower lip rounded rather than cuneate, lateral lobes rounded, not acuminate. Winter leaves are larger than summer ones and hair cover thinner to promote photosynthesis.

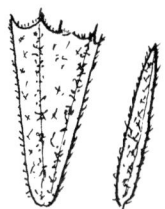

Shrub to 90 cm. Leaves oblong-lanceolate, obtuse, attenuate to a short petiole, rarely subcordate, thick, reticulate veined, stellate-tomentose on both surfaces, slightly crenate to entire at margin, greenish yellow, young ones golden. Floral leaves sessile. Whorls distant, few flowered. Bracteoles oblong to linear, almost the length of the calyx. Calyx tubular c. 15 cm, yellow stellate-tomentose with short triangular, subulate equal teeth. Corolla yellow. Nutlets hairless.

Phlomis bourgaei Boiss.

Fl. Or. 4:787 (1879)
Synonyms:
    P. viscosa Poir. subsp. *bourgaei* (Boiss.) P.H.Davis
    P. schwarzii P.H.Davis
Distribution in the wild:
    Endemic to Turkey, in maquis, *Quercus* scrub, *Pinus* woods,
    calcareous and serpentine rocks from sea level–1000 m.
Flowering in the wild: April–August
bourgaei is in honour of Eugène Bourgaeu (1813-1877).

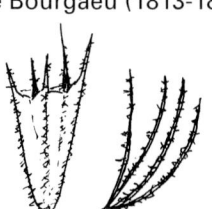

Differs from *P. viscosa* Poir. by having narrower bracteoles, a smaller calyx (with longer teeth), by having a smaller denser inflorescence and has less glandular hairs.

Evergreen shrub to 150 cm. Leaves glandular-stellate above and densely dendroid-tomentose beneath, triangular-ovate to oblong-lanceolate, cordate at base, crenate-serrate at margin, 3–16 × 1.5–6 cm; petiole to 5 cm, edged in glandular dendroid hairs. Floral leaves triangular-ovate-lanceolate; shortly petiolate. One to two whorls, (6–)12–20 flowered. Bracteoles subulate, densely long hispid-viscid 12–17 × 0.75–1.5 mm, teeth subulate, 3–7 mm. Corolla yellow, 20–30 mm. Nutlets hairless.

Phlomis bourgaei 'Whirling Dervish' ✦❀

Colour Plate: VII (i)

A selection from the wild with very cockled leaves was introduced by the author. Hardy to –15°C.

Natural hybrids:

P. × mobullensis Hub.–Mor.

Bauhinia 6(3):373 (1979)

(= *P. bourgaei* Boiss. × *P. grandiflora* H.S.Thompson)

Distribution: Turkey

Differs from *P. bourgaei* in having bracteoles 2-3 mm wide and no glandular hairs on the calyx teeth.

Differs from *P. grandiflora* by having glandular hairs on leaves and bracteoles 2-3 mm wide.

P. × termessi P.H.Davis

Kew Bull. 1951:93 (1951)

(= *P. bourgaei* Boiss. × *P. lycia* D.Don)

Distribution: Turkey

Phlomis brevibracteata Turrill

*Gard. Chron. Ser.*3, 117:48 (1945)

Distribution in the wild:

Endemic to Cyprus (north & south), on garigue, on dry rocky forest soil, together with *Cistus*.

brevibracteata means with short bracteoles.

Flowering in the wild: May–June

Distinguished from other Cyprus species by its small, narrow bracteoles.

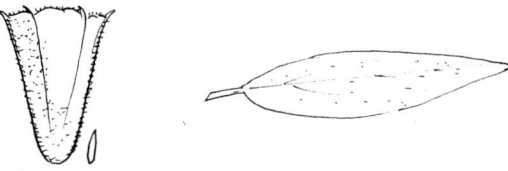

Evergreen shrub to 150 cm. Lax, spreading habit. Leaves oblong, bright green above, white tomentose below, apex obtuse, or rounded, or broadly cuneate at base, 2–5 × 0.7–1.5 cm; petiole 0.4–0.8 cm. Floral leaves very similar to leaves. Two to three whorls per stem, 4–12 flowered. Bracteoles very small, linear-subulate, 2–5 × 1 mm densely greyish tomentose. Calyx tubular, 10–12 mm, 7–8 mm wide at mouth, densely stellate-tomentose. 10 veins; teeth inconspicuous. Corolla rich yellow. Nutlets hairless.

Phlomis chimerae Boiss.

Bull. Soc. Bot. Fr. 43:290 (1986)
Distribution in the wild:
   Endemic to Turkey, in *Pinus halepensis* (syn: *brutia*) forest, maquis,
   rocky slopes, sea level to 150 m.
Flowering in the wild: April–August.
chimerae is named after the mythological fire breathing monster, the
Chimaera, which ravished Caria and Lycia in Asia Minor, where the plant is
found.
A **rare** species in the wild.

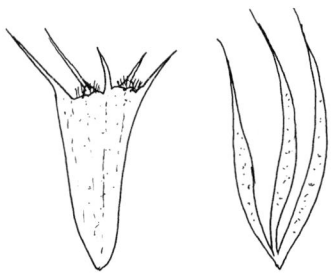

Non-glandular dwarf shrub to 30 cm. Leaves greenish with minute stellate hairs and often
glabrescent above, densely white-tomentose below, longer undivided hairs absent. Cauline
leaves small, oblong or ovate to broadly ovate, obtuse, shortly cuneate, rounded, truncate
or subcordate at base, entire or indistinctly crenulate at margin, 1.5–4 × 1–2.5 cm; petiole to
2 cm. Floral leaves shortly petiolate, oblong. One to two whorls, 6–12 flowered. Bracteoles
linear-subulate, white stellate-tomentose, 13–17 × 1–1.5 mm, curved upwards. Calyx 15-20
mm, densely white-tomentose, teeth spreading, unequal, 3–8 mm. Corolla yellow, 25–30
mm. Nutlets hairless.

Phlomis chrysophylla Boiss. ✦✧

Diagn. ser. 1,12:89 (1853)
Colour Plate: II (i)
Distribution in the wild:
   Lebanon, Syria and Jordan
Flowering: May–July in the wild.

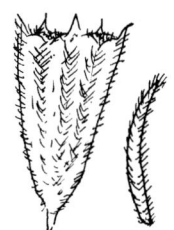 

chrysophylla means with golden leaves and refers mainly to dried herbarium specimens although some live material is also yellowish (slightly khaki), developing at different seasons.

Evergreen shrub to 1m. Lower leaf laminas broadly elliptic to oblong-ovate, obtuse to rounded at apex, truncate or rounded at base to 8.5 × 6 cm; petiole to 5 cm. Floral leaves 2.8–6.8 × 2–5.3 cm, circular to obovate, petiole to 1.5 cm, edged with stellate and long multijointed hairs. One to two whorls, many flowered; occasionally in pairs when calyces are often stalked to 7 mm. Whorls 6–7 cm across. Bracteoles linear subulate, ciliate to 12 × 1 mm often in groups of three joined at base. Calyx to 15 mm; teeth broad and short with a mucro of 1 to 2 mm. Corolla 28–32 mm, golden yellow. Nutlets hairless. Hardy to –15°C.

Phlomis cretica C. Presl

Del. Prag. 84 (1822)
Colour Plate: III (ii)
Synonym:
   P. ferruginea var. *cretica* Benth.
Distribution in the wild:
   Crete, Greece, Aegean Islands in clearings of *Quercus coccifera* woodland, garigue. Sea level to 1,250m.
Flowering in the wild: March–May
cretica means of Crete.

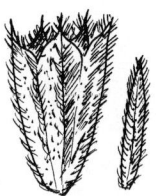

A compact, low growing attractive shrub with grey felted leaves often with a yellow tinge to the lower surface and edges. Hardy to –10°C.

Dwarf evergreen shrub to 45 cm. Leaves stalked stellate-tomentose above, stalked stellate-lanate beneath. Lower leaves lanceolate to oblong, crenulate, cuneate, truncate or rounded at base, 3–19 × 1.5–6.5 cm; petiole to 4 cm. Single whorls 14–30 flowered. Bracteoles subulate, linear or narrowly lanceolate, stellate-tomentose and ciliate, 12–19 × 0.75–3 mm. Calyx 13–19 mm stellate-lanate with glandular hairs. Teeth subulate 1–4 mm. Corolla yellow, 25–27 mm. Nutlets hairless.

Natural hybrids:
   P. × commixta Rech. f.
Denkschr. Akad. Wiss. Wien. 105(2,1):119 (1943)
(= *P. cretica* C. Presl × *P. lanata* Willd.)
Distribution: Crete, in clearings of *Quercus coccifera* woodland, at 800m.
Leaves 2–3 cm long. Bracteoles 12–15 × 2–3 mm. Calyx 15 mm; teeth mucros 1–1.5 mm.
   P. × cytherea Rech. f.
Boissiera 13:115 (1967)

(= *P. cretica* C. Presl × *P. fruticosa* L.)
Distribution: Greece, Crete at 300–500 m.
Flowering: April–May in wild.
Leaves oblong-ovate. Bracteoles lanceolate to 4 mm wide. Calyx teeth horizontal, mucro to 4 mm with glandular hairs.

Phlomis cypria Post ◆✧

Mém. Herb. Boissier 18:99 (1900)
Synonym:
    P. fruticosa L. subsp. *cypria* (Post) H. Lindb. f.
Distribution in the wild:
    Endemic to Cyprus
Flowering in the wild: April–June
cypria means of Cyprus.

Evergreen shrub 50–150 cm. Leaves ovate-oblong or oblong-lanceolate, dull green above, shortly stellate; greyish or yellowish stellate tomentose below, 2–7 × 0.5–3 cm; petiole 0.5–2.5 cm. Floral leaves subsessile and smaller. One or two whorls, many flowered. Bracteoles adpressed, obovate or rhomboid-obovate 7–12 × 3–7 mm. Calyx tubular, 13–15 mm, 6–8 mm wide at apex, hairless except for a dense zone of yellowish or whitish hairs at the mouth. 10 veins; teeth 2–2.5 mm. Corolla pale yellow. Nutlets glandular-papillose at apex.

P. cypria Post var. *cypria*

Grows on sunny limestone hillsides

Leaves shortly and broadly oblong, 3 – 4 × 1.5 – 2 cm, whitish or greyish tomentose, with a rounded or very obtuse, broad apex.

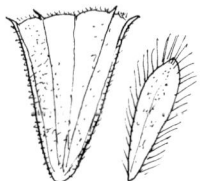 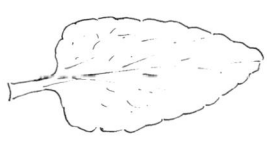

P. cypria Post var. *occidentalis* Meikle ◆✧

Ann. Musei Goulandris 6:93 (1983)
Grows only on igneous & serpentine hillsides in the south of the island.
Colour Plate: IV (iii)

Leaves narrowly oblong, 3 – 7 × 1 – 2.3 cm, often yellowish tomentose, tapering to a narrow, obtuse or subacute apex. Hardy to –10°C.

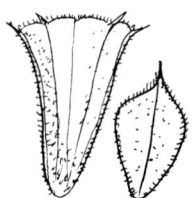 

Phlomis 'Edward Bowles'

Raised and named by Hilliers in 1967, allegedly from seed of E. A. Bowles (who however, died in 1954). Has been said to be a hybrid between *P. russeliana* (perennial) and *P. fruticosa* (shrub), but at the moment its provenance is 'not proven'.

Colour Plates: IV (i), (ii)

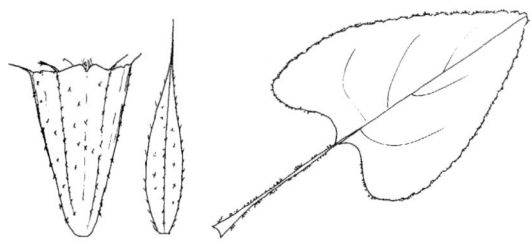

Shrub to 1½ m. All parts with stellate hairs. Leaves bright green above, silvery white beneath, elongated heart-shaped, narrowly rimmed in white from stellate hairs beneath, to 20 × 11 cm; petiole to 12 cm. Floral leaves lanceolate to ovate, 5.8–11.7 × 2.1–8.5 cm, petiole to 1 cm. Flowering stems bearing up to 4 whorls per stem. Whorls 7 – 8 cm across. Bracteoles 16–22 × 3–6 mm, outer ones covering smaller inner ones, acuminate. Calyx to 18 mm, teeth to 5 mm, held horizontally. Corolla, 27–32 mm, two toned, upper lip cream, lower lip golden yellow. Central lobe of lower lip to 20 mm across. Nutlets hairless. Hardy to –15°C

Phlomis elliptica Benth.

Lab. Gen. et Sp. 626 (1834)

Distribution in the wild:

Endemic to Iran

elliptica means elliptic and refers to the shape of the leaves.

Shrub 40–50 cm. Leaves elliptic, cuneate at base, apex obtuse, green above, white stellate-tomentose below, 4.5 × 2 cm; petiole to 1 cm. Usually one terminal whorl, many flowered. Bracteoles subulate, stellate-tomentose. Calyx 12–13 mm, narrowly funnel shaped; teeth broad based with mucro to 2 mm. Corolla rose-coloured.

Phlomis ferruginea Ten.

Fl. Napol. Prodr. 1,35:10 (1811-1815)

Synonyms:

P. viscosa Parl.

P. viscosa var. *ferruginea* Bég.

Distribution in the wild:

Endemic to S. Italy, on garigue, sea level–300 m.

ferruginea means rust-coloured and refers to the stems.

It has been reported from Sinai and Crete, but incorrectly. Said to be a little bit like a small *P. fruticosa* with rust coloured stems. **Rare** in the wild.

Evergreen shrub 40 –100 cm. Whorls 12–16 flowered. Bracteoles linear, ciliate, 8–10 mm. Calyx with teeth 1–3 mm.

Phlomis floccosa D.Don

Bot. Reg. 15:t.1300 (1830)
Synonyms:
> *P. bicolor* (Viv.) Benth.
> *P. lanata* Gand.
> *P. samia* var. *bicolor* Viv.

Distribution in the wild:
> Tunisia, Libya, Egypt, Karpathos & Kos, in scrub, garigue, sea level–
> 900 m.

Flowering in the wild: March–April
floccosa means woolly, looking like matted wool.
Only one characteristic is really needed to define this plant and that is the hooked calyx teeth. Tender.

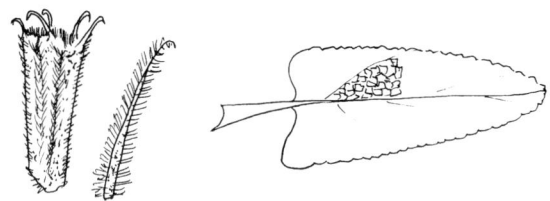

Shrub to 1m. Lower leaves oblong-ovate or ovate-lanceolate, cordate or sub-cordate at base, crenate at margin, rugose on upper surface, stellate-tomentose on both surfaces, 3–5 cm long; petiole to 3 cm. Floral leaves lanceolate, acute or acuminate; shortly petiolate. Whorls 4–8 flowered. Bracteoles numerous, linear-subulate, 15–18 mm, stellate-lanate, ciliate with 2–3 mm hairs. Calyx tubular, 15–19 mm, stellate-lanate, ciliate; calyx teeth subulate, unciniate, 1+5 mm. Corolla 25–32 mm, yellow. Nutlets glabrous.
> Natural hybrid:
> > ***P. × vierhapperi*** Rech. f.
> > *Ann. Naturhist. Mus. Wien* 47:149 (1936)
> > (= *P. floccosa* D.Don × *P. pichleri* Vierh.)
> > Distribution: Karpathos

Phlomis fruticosa L. (Common name: Jerusalem Sage)

Sp. Pl. 584 (1753)

Illust.: *Bot. Mag.* t.1843 (1816)
Synonyms:
> *P. latifolia* Mill.
> *P. portae* Kerner ex Nyman

Distribution in the wild:
> Sardinia, Malta, Sicily, Italy, former Yugoslavia, Albania, Greece, Crete
> & Karpathos, Cyprus, Turkey and the former USSR, on rocky
> limestone slopes, scrub, garigue at sea level–1000 m.

Flowering in the wild: April–July
fruticosa means shrubby.

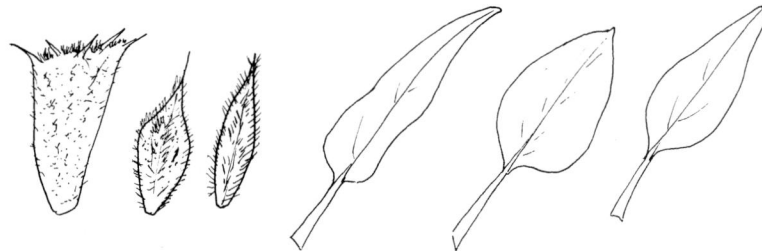

In cultivation since 1596. Although commonly called Jerusalem Sage, the shrub is not found in Israel and Jordan.

A very variable shrub with many forms of leaf shape and bracteole width. Can grow to a great size and be of considerable age. Responds well to pruning and will regenerate from hardwood.

P. fruticosa 'Butterfly' ✧ introduced in France.

P. fruticosa 'Compacta' ✧ (invalid) introduced in France.

P. fruticosa 'Crispy' ✧ introduced in France has undulating leaf edges.

P. fruticosa 'Speckles' ✦ is a variegated form which occurred in a Gloucestershire garden.

Evergreen shrub to 40 cm–2 m, more in width. Grey woolly stellate. Lower leaf laminas elliptic, lanceolate-ovate or lanceolate, truncate or cuneate at base, entire or crenulate at margin, 3–9 × 1.6–3.5 cm; petiole to 4 cm. Floral leaves lanceolate to ovate, greenish, 4–6.5 × 1.8–5.5 cm, almost stalkless. Flower stems white felted, with 1–2 whorls, many flowered. Whorls 6–7 cm across. Bracteoles obovate or broadly lanceolate, acuminate, 10–20 × 2–7 mm, adpressed to calyces, with numerous short stellate and longer undivided weak hairs. Calyx 10–20 mm densely stellate-lanate, not ciliate; teeth 1–4 mm held horizontally. Corolla 23–35 mm, yellow to orange. Nutlets hairless or hairy. Hardy to–15°C.

Natural hybrids:

P. x cytherea Rech. f.

Boissiera 13:115 (1967)

(= *P. cretica* C. Presl × *P. fruticosa* L.)

Distribution: Greece

Leaves oblong-ovate. Bracteoles lanceolate to 4 mm wide. Calyx teeth horizontal, mucro to 4 mm with glandular hairs.

P. x sieberi Vierh.

Öst. Bot. Zeit. 65:231 (1915)

(= *P. fruticosa* L. × *P. lanata* Willd.)

Distribution: Crete

Phlomis grandiflora H.S.Thompson   ✦

Ann. Bot. 19:441 (1905)

Colour Plate: II (ii)

Synonyms:

P. lunariaefolia sensu Boiss.

P. lunarifolia Benth.

Distribution in the wild:

Greece and Turkey, in *Pinus halepensis* (syn: *brutia*) forest, *Quercus* scrub, maquis, limestone slopes, rocks at 600–1220 m.

Flowering in the wild:　　May–August

grandiflora means large-flowered.

Although this does have very large showy flowers they are only produced in profusion occasionally (usually following a very hot previous summer) and on rather long ungainly stems. The lack of flowering however allows the plant to become very useful as a handsome foliage plant. It has large grey leaves that tend to be folded upwards along the central vein. Very old leaves loose their indumentum and become large and greenish.

Evergreen shrub to 2 m. adpressed stellate hairy. Older leaves green above, new leaves grey felted on upper surface with stellate and stalked stellate hairs, ovate to oblong, shortly cuneate, rounded, truncate or cordate at base, entire or crenulate at margin, 3–8 × 2–4 cm; petiole to 8 cm with dendroid hairs. Young grey leaves folded along central vein. Floral leaves lanceolate, 6.5–8 × 3–3.5 cm, stalkless or with short petiole. White felted stems with single (occasionally 2) whorls of many flowers. Whorls to 10 cm across. Bracteoles numerous, broadly ovate to ovate-lanceolate, acuminate. 12–18 × 3–10 mm, adpressed, hairless above, sparsely stellate below. Calyx densely stellate plus less numerous undivided hairs, 13–17 mm; teeth 2–3 mm held horizontally. Corolla 30–40 mm, yellow. Nutlets hairless. Hardy to –15°C.

P. 'Lloyd's Variety'　✦✧　A form of *P. grandiflora* H.S.Thompson

P. 'Anatolica' (*P. anatolica* invalid)　✦✧ A form of *P. grandiflora* H.S.Thompson

P. grandiflora 'Suleiman'　✦✧ introduced by the author, but withdrawn due to lack of distinguishing features.

P. grandiflora H.S.Thompson **var. grandiflora**　✦✧

Synonyms:

P. lunariaefolia sensu Boiss.
P. lunarifolia Benth.

Lower leaves rounded or truncate at the base.

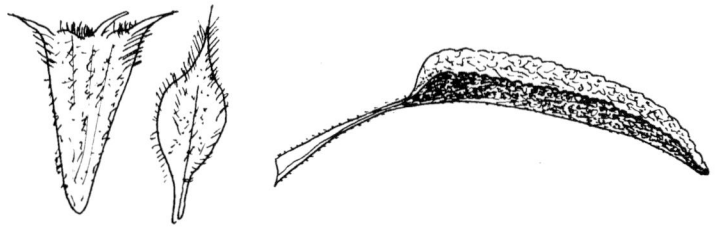

Natural hybrids:
P. × muglensis Hub.–Mor.
Bauhinia 6(3):374 (1979)
(= *P. grandiflora* H.S.Thompson × *P. lycia* D.Don)
Distribution: Turkey
Differs from *P. grandiflora* with 2 whorls, calyx teeth 1–2 mm long and bracteoles 2–5 mm wide.

Differs from *P. lycia* with 2 whorls, wider leaves to 3 cm, calyx to 15 mm, teeth to 2 mm and bracteoles to 5 mm wide.

P. x mobullensis Hub.–Mor.

Bauhinia 6(3):373 (1979)

(= *P. bourgaei* Boiss. × *P. grandiflora* H.S.Thompson)

Distribution: Turkey

Differs from *P. bourgaei* in having bracteoles 2-3 mm wide and no glandular hairs on the calyx teeth.

Differs from *P. grandiflora* by having glandular hairs on leaves and bracteoles 2-3 mm wide.

P. grandiflora H.S.Thompson **var. fimbrilligera** Hub.–Mor.

Bauhinia 1(20:107 (1958)

Synonym:

P. fimbrilligera Hub.–Mor.

Distribution in the wild: Turkey

Found nearer the sea some 200km south-east of the main *P. grandiflora* area. Considered a splendid and floriferous plant by Huber-Morath.

In this variety the bracteoles are more strongly fringed with hairs, and it has smaller, more closely and regularly crenate leaf laminas, which with the exception of the upper leaves on flowering stems, are always clearly cordate at the base.

Phlomis italica L.   ✦✧❀

Syst. Nat. ed. 10 (2):1102 (1759)

Illust.: *Bot. Mag.* t. 9270 (1929)

Colour Plate: IV (iv)

Synonyms:

P. rotundifolia Mill.

P. balearica Chodat

P. italica Smith

The epithet *italica* was given by Linnaeus as it had arrived from or via Italy. Chodat's epithet *balearica* of 1905, if it had been accepted, would have saved a lot of arguments, as *Phlomis italica* is endemic to the Balearic Islands particularly the Gimnesias ones (Mallorca and Minorca).

Originally described as a sub-shrub up to about 30 cm (*Botanical Magazine* t.9270) and described as such in the *Index of Garden Plants* by Mark Griffiths (1994), this was revised to 2 m (Mateu, 1986). The taller shrubs are more numerous in Mallorca. The smaller growing forms in cultivation have an inferior flower colour (lower lip deep blue/ purple), but only grow to about 1 metre.

Recognised from the lack of visible calyx teeth and the shorter bracteoles, both covered in a dense loose indumentum.

A popular plant in our gardens because of its hardiness and the pleasant pink colour of one of the clones in cultivation. This pink form grows up to 2 m. in our gardens, but gradually becomes somewhat naked in the lower regions. Better cut back regularly, or best replaced after five years if not. Although it has

been in cultivation for many years I have now given it the name ***P. italica***
'Pink Glory' ✦❀ to distinguish it from the inferior smaller ones now being introduced.
Grown by Philip Miller (1768) at the Chelsea Physic Garden.

P. italica L. **subsp.** ***italica***   ✦✧❀
Distribution in the wild:
    Endemic to the Balearic Islands.

Evergreen shrub to 2 m, commonly 1 m. Lower leaf laminas ovate-lanceolate, densely stalked stellate woolly, cordate or truncate at base, crenate at margin, 3 – 9 × 1.5 – 3 cm; petioles to 6 cm. Floral leaves lanceolate-oblong 3–6 × 0.8–2 cm; petiole to 1.2 cm. White felted stems (easily removable stellate hairs) bearing up to 4 whorls of around 10 flowers. Whorls 3–4 cm across. Bracteoles to 7–12 × 1.5 mm. Calyx tubular to 15 mm, rounded equal teeth to 2 mm with a minute mucro usually hidden in the indumentum (visible in late winter when the indumentum wears off). Bracteoles and calyx densely felted with loosely held stellate hairs. Corolla to 20 mm, colour variable, upper lip often pink to white, lower lip sometimes blue/purple or white. Nutlets hairless or with very short glandular hairs. Hardy to −15°C.

P. italica L. **subsp.** ***antiatlantica*** (Peltier) Rivas Mart.
Acta Bot. Malacitana 2:59-64 (1976)
Distribution in the wild:
    Morocco at around 1500 m.
This subspecies is found in the Anti-Atlas mountains of Morocco and has a funnel shaped calyx 11-13 mm; corolla 17-20 mm, pink; rounded teeth with teeth to 0.3 mm, covered in stellate-tomentose indumentum. The nutlets have stellate hairs on the apex.

Phlomis lanata Willd.   ✦✧

Enum. Pl. Hort. Berol. Suppl. 41 (1814)
Colour Plate: V (iv)
Synonyms:
    P. parvifolia C. Presl
    P. microphylla Sieber
Distribution in the wild:
    Crete & Karpathos, in rocky clearings of *Cupressus* woodland, garigue, cliffs, sea level to 1,600 m.

Flowering in the wild: April–May
lanata means woolly and refers to the stems and calyx.
Easily characterised by its small grey elliptical or roundish leaves.

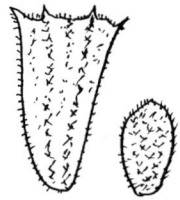 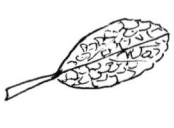

Evergreen shrub to 60 cm. Lower leaf laminas broadly elliptic, oblong, obovate or almost circular, wrinkled shortly stalked stellate, 1.5–4.5 × 1.1–2.7 cm; petiole 0.5–2 cm. Floral leaves almost circular, obtuse, 2.3–3.5 × 1.3 –2 cm; petiole to 0.5 cm. White or yellow felted stems with 1–2 whorls and up to 10 flowers per whorl. Whorls 5–6 cm across. Bracteoles broadly elliptical, oblanceolate or obovate, mucronate, acuminate, 10–20 × (2–)3–7 mm. Calyx 10–15 mm with teeth to 1 mm. Bracteoles and calyx densely felted with stellate hairs. Corolla, 20–32 mm, yellow, often with a brownish tint to the crest of the upper lip. Nutlets hairy. Hardy to –15°C

A form (ultimate size not yet known) introduced in France, with even smaller leaves is also in cultivation and has been given the name *P. lanata* 'Pygmy' (*P. lanata* 'Nana' invalid)

> Natural hybrid:
> ## *P.* x *commixta* Rech. f.
> *Denkschr. Akad. Wiss. Wien. Math.-Nat.* Kl2.Abt.1,119 (1943)
> (= *P. cretica* C. Presl × *P. lanata* Willd.)
> Distribution: Crete, in clearings of *Quercus coccifera* woodland at 800 m.
> Leaves 2–3 cm long. Bracteoles 12–15 × 2–3 mm. Calyx 15 mm; teeth mucros 1–1.5 mm.
> ## *P.* x *sieberi* Vierh.
> *Öst. Bot. Zeitschr.* 65:231 (1915)
> (= P. *fruticosa* L. × *P. lanata* Willd.)
> Distribution: Crete

Phlomis leucophracta P.H.Davis & Hub.–Mor.   ◆✧❀

Kew Bull. :90 (1951)
Colour Plates: III (i), (iii), (v)
Distribution in the wild:
        Endemic to Turkey, on limestone rocks, metamorphic slopes, maquis, *Quercus* scrub, fallow fields at sea level–1000 m.
Flowering in the wild: June–August
leucophracta means white edged, referring to the leaf margins.
This beautiful shrub is easily recognisable from its leaves which are green above and edged in a 'blanket stitch' rim of hairs which can be white to yellow. The upper lip of the corolla is a rust-brown colour. In plants under stress, the edges of the leaves roll over the top surface to meet in the centre, ensuring

reduced transpiration.
There are both tender and hardy forms in cultivation.

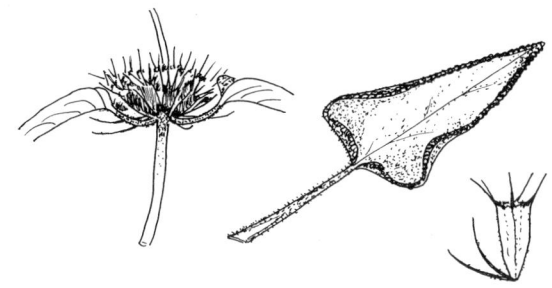

Evergreen shrub to 1.5 m. Lower leaf laminas triangular-lanceolate-ovate, cordate at base, crenate at margin, 5–12 × 2–5 cm; petiole to 4.5 cm. Upper surface of leaf bright green with a broad edge of white or golden hairs protruding from the underside. Densely dendroid beneath. Floral leaves oblong-lanceolate, 4–5.5 × 2.5–3.5 cm; petiole 1–3 cm with long multicellular hairs. Felted stems, bearing 1–3 whorls of up to 12 flowers. Whorls 6–8 cm across. Bracteoles subulate, narrowly lanceolate, 15–22 × 1–2 mm, covered with sticky glandular stellate hairs. Calyx 20–27 mm, sticky with subulate, erect teeth, short ones 5–9 mm, long ones 10–12 mm. Bracteoles and calyces widely divergent. Corolla 30–35 mm, upper lip of corolla rust brown, lower lip yellow or outer lobes rust coloured. Nutlets hairless.

A selected wild-collected form with a strong yellow indumentum on the stems and leaves was introduced by the author *P. leucophracta* **'Golden Janissary'** ✦✧❀ Hardy to –15°C Colour Plate: III (iii)
P. leucophracta **'Silver Janissary'** ✦ is similar to *P. leucophracta* 'Golden Janissary', but with white hairs and therefore a white edge to the leaf. Hardy to –10°C. Colour Plate: III (i)
P. leucophracta **'White Edge'** ✧ is similar to *P. leucophracta* 'Silver Janissary' but introduced in France.
Natural hybrids:
P. x alanyense Hub.–Mor.
Bauhinia 1(2):111 (1958)
(= *P. leucophracta* P.H.Davis × *P. lunariifolia* Smith)
Distribution: Turkey
A natural Turkish hybrid between *P. lycia* and *P. leucophracta* has been tentatively named *Phlomis* **'Goldmine'** ✦❀❀ It has the leaf shape of *P. lycia* with a golden indumentum and brown upper lip of *P. leucophracta*. Hardy to -15°C.

Phlomis longifolia Boiss. & Blanche ✦✧

Diagn. ser. 2(4):47 (1859)
Colour Plate: VII (iii)
Distribution in the wild:
Cyprus?, Lebanon, Turkey, on limestone slopes and rocks, maquis, *Quercus* scrub at 1600–1050 m.

Flowering in the wild: May–June

longifolia indicates that it has long leaves.
Some of the forms in cultivation have a very rich deep green colour with a pronounced puckered (rugose) upper surface, especially on the floral leaves. Their flower colour is an egg yolk yellow. In the wild I have found it very difficult to find one corresponding to the subspecies *bailanica*, even in the *locus classicus* of Belen in Turkey. The leaf shape and calyx teeth are quite variable. Distinguished from *P. viscosa* Poir. by the absence of glandular hairs and stiffer hairs on the bracteoles and calyx.

Non-glandular evergreen shrub to 1.3 m. Green lower leaf laminas, upper surface with stellate and simple hairs, lower surface stalked stellate, lanceolate to oblong or ovate, puckered, cordate or sub-cordate at base, crenulate or crenate-serrate at margin, 3–10 × 1.5–4 cm; petiole to 5 cm with stalked stellate hairs. Floral leaves lanceolate to ovate-triangular, 4.5–10 × 2.2–5 cm, petiole to 1 cm. One to three whorls per stem, to 20 flowered. Whorls 5–7 cm across. Bracteoles numerous, linear-subulate to linear-lanceolate 8–20 × 1–3 mm, densely bristly. Calyx 15–20 mm, stellate hairy. Teeth 1 to 6 mm and held horizontally. Corolla 30–40 mm, golden yellow. Nutlets hairless. Hardy to –15°C.

P. longifolia Boiss. & Blanche **var. *longifolia***
Synonyms:
> *P. parvifolia* Post
> *P. viscosa* Poir. var. *angustifolia* Boiss.
> *P. bertrami* Post

Basal leaf laminas 3-5 times as long as broad; calyx teeth 1 – 3 mm.

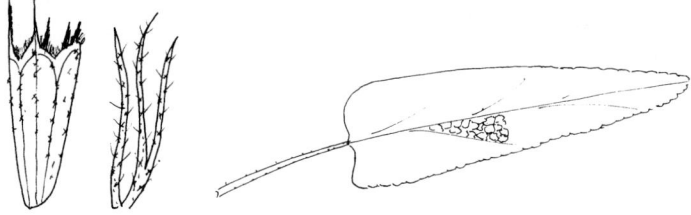

P. longifolia var. *bailanica* (Vierh.) Hub.–Mor.
Bauhinia 1(2):112 (1958)
Synonym:
> *P. bailanica* Vierh.

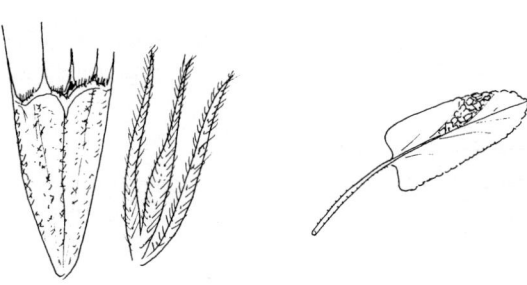

Distribution in the wild:
Turkey, Syria
A broad leafed form, looking more like *P. viscosa* Poir., but without the glandular hairs.
Basal leaf laminas 2 times as long as broad; calyx teeth 3 – 6 mm.
Most forms labelled *P. longifolia* var. *bailanica* in cultivation are probably *P. longifolia* var. *longifolia*.

Phlomis lunariifolia Smith

Fl. Graec. Prodr. 1:414 (1809)
Colour Plate: VIII (i)
Synonym:
P. imbricata Boiss.
Distribution in the wild:
Cyprus, Turkey, in *Pinus halepensis* (syn: *brutia*) woods, maquis, rocky limestone slopes from sea level–640 m.
Flowering in the wild: May–July
lunariifolia means having crescent shaped leaves and probably refers to the way the leaves and floral leaves hang.
Often found in the wild together with *P. leucophracta* P.H.Davis, this plant is related to *P. grandiflora* H.S.Thompson.

Evergreen shrub to 130 cm. Leaves shortly stellate-hairy, greenish above, white-stellate tomentose below, lower leaves ovate to lanceolate, shortly cuneate, rounded or truncate at base, entire to crenate-serrate at margin, 4–8 × 1–3 cm; petiole to 3 cm. Floral leaves sessile or with a short petiole, lanceolate to linear-lanceolate. One to two whorls per stem (6)12–28 flowered. Bracteoles narrowly lanceolate to lanceolate, prickly acuminate, 7–12 × 2–3 mm, sparsely stellate hairy to almost hairless, with marginal stellate and rigid hairs. Calyx 12–18 mm, ± stellate-tomentose above, hairless below; teeth subulate, 2–3 mm. Corolla yellow, 20–30 mm. Nutlets hairless. Hardy to –15°C
Natural hybrids:
P. × cilicica Hub.–Mor.
Bauhinia 1(2):113 (1958)
(= *P. lunarifolia* Smith × *P. monocephala* P.H.Davis)
P. × alanyense Hub.–Mor.
Bauhinia 1(2):111 (1958)
(= *P. leucophracta* P.H.Davis × *P. lunariifolia* Smith)
Distribution: Turkey

Phlomis lycia D.Don ✦ ✧

Discov. Lycia 293 (1841)
Distribution in the wild:
    Endemic to Turkey, in maquis, *Quercus* shrub, *Pinus halepensis* (syn:
    brutia) forest, serpentine cliffs at sea level–640 m.
Flowering in the wild: May–August
lycia means of Lycia in SW Turkey.
Some forms in cultivation are less hardy than others.

 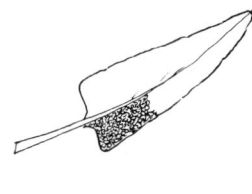

Non-glandular evergreen shrub to 1.5 m, Lower leaf laminas oblong-lanceolate, dense
stellate hairy (white or yellow), with or without cordate base, crenulate at margin, 2–10 ×
0.7–.3 cm; petiole 1–4.5 cm. Floral leaves often larger and broader, cordate at base, with a
short petiole. Felted stems (white or yellow) bearing 1–2 whorls per stem, each with up to
12 flowers. Bracteoles linear-lanceolate, densely lanate, 8–11 × 1–2 mm. Calyx 10–12 mm
densely lanate, teeth to 1 mm. Corolla 25–30 cm, yellow. Nutlets hairless. Hardy to –10°C.

A natural hybrid between *P. lycia* and *P. leucophracta* has been
tentatively named ***Phlomis* 'Goldmine'** ✦ ❀ ❀   It has the leaf
shape of *P. lycia* with a golden indumentum and brown upper lip of *P.
leucophracta*. Hardy to –15°C.

    Natural hybrid:
        P. × muglensis Hub.–Mor.
    Bauhinia 6(3):374 (1979)
    (= *P. grandiflora* H.S.Thompson × *P. lycia* D.Don)
    Distribution: Turkey
    Differs from *P. grandiflora* with 2 whorls, calyx teeth 1–2 mm long
    and bracteoles 2–5 mm wide.
    Differs from *P. lycia* with wider leaves to 3 cm, calyx to 15 mm,
    teeth to 2 mm and bracteoles to 5 mm wide.
        P. × termessi P.H.Davis
    Kew Bull. 1951:93 (1951)
    (= *P. bourgaei* Boiss. × *P. lycia* D.Don)
    Distribution: Turkey (Termessus) at 600 m.

Phlomis monocephala P.H.Davis   ✦ ✧

Kew Bull. 4:412 (1949)
Distribution in the wild:
    Endemic to Turkey, in maquis, *Quercus* & *Cedrus* scrub, rocky
    calcareous woods at sea level–1500 m.
Flowering in the wild: May–August

monocephala indicates only one flower head per stem, although it does occasionally have two per stem.

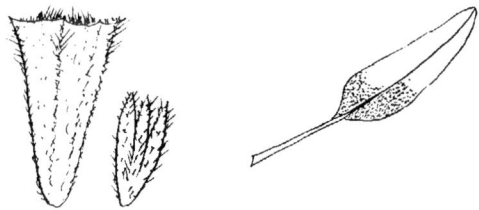

Non-glandular evergreen shrub to 1.50 m. Leaves densely stellate tomentose, longer undivided hairs absent, lower cauline leaves oblong to oblong-ovate, obtuse, cuneate at the base, denticulate or entire at margin, 2–6.5 × 1–3.5 cm; petiole to 3 cm. Floral leaves with almost no petiole, oblong. One to two whorls per stem with 6–12 flowers. Bracteoles lanceolate, densely spreading white-lanate, 5–8 × 1.5–2 mm, Calyx 10–14 mm, densely white lanate; teeth broadly triangular, scarcely apparent with mucro 0.5 mm. Corolla yellow, 20–30 mm. Nutlets hairless.

Natural hybrid:

P. × cilicica Hub.–Mor.

Bauhinia 1(2):113 (1958)

(= *P. lunariifolia* Smith × *P. monocephala* P.H.Davis)

Phlomis pichleri Vierh.

Öst. Bot. Zeitsch. 65:232 (1915)

Synonym:

P. ferruginea Barb.

Distribution in the wild:

Endemic to Karpathos group, in scrub, garigue, rocky gorge-beds, sea level to 800 m.

Flowering in the wild: April to May

pichleri is after R. Alfred Pichler.

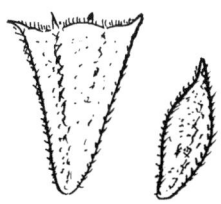

Leaves oblong-ovate, rugose, green above, tomentose beneath, two to three times as long as wide, broadly rounded–deeply cordate at base, obtuse at apex. Whorls many flowered. Bracteoles imbricate, oblong-ovate–broadly lanceolate, stellate-tomentose. Calyx tubular, stellate-tomentose, 14 mm; teeth angled 1–2 mm. Corolla yellow, to 32 mm.

Phlomis platystegia Post—see page 54

Phlomis purpurea L. ✦✧

Sp. Pl. 585 (1753)
Illust.: *Bot. Mag.* t. 518 (1968)
Colour Plates: VI (iii), (iv), (v), (vi)
Synonym:
   P. salviaefolia Jacq.
Distribution in the wild:
   S Portugal, S Spain and Morocco.
purpurea means purple and refers to the most common corolla colour.
This shrub is best characterised from its whorl and bracteoles. The more
compact plants in cultivation seem to suffer from bad suckering and those that
don't sucker tend to be rather lax in growing, benefiting from being grown
amongst other shrubs for support. Careful selection from the wild of a good
growing form with a good pink colour is a priority. The species is very variable
in the wild, with leaves from green to grey and flower colour from deep purple
through pink to white. Some of the forms distributed in commerce seem to
suffer tip damage from sub-zero temperatures in our gardens. Greatly benefits
from being cut back regularly.

Branching lax evergreen shrub to 2 m. high (more commonly around 1 m.). Lower leaf
laminae green through grey to khaki, ovate, lanceolate or oblong with truncate or cordate
base, entire or crenate at margin, stalked stellate indumentum, 4–10(–14) × 0.7–5(–6.5) cm;
petiole to 5cm. Floral leaves greener than lower leaves , lanceolate, 4–7 cm by 1–2 cm;
petiole 0.5–1.2 cm. Flowering stems simple or branched, bearing up to 5 whorls per stem,
each with up to 12 flowers. Whorls 4–5 cm across. Bracteoles lanceolate or elliptic,
acuminate, 10–18 × 3–5 mm, adpressed. Calyx to 16 mm, 10 veins, with teeth to 6 mm and
a mucro 2–4 mm. Corolla 23–26 mm, purple with lower lip sometimes almost white. Nutlets
hairless or with very short glandular hairs. Hardy to –10°C.

P. purpurea L. subsp. *purpurea* ✦✧

Colour Plates: VI (iii), (v), (vi)
Leaves 5–10(-14) × 1–5(-6.5) cm. Mucro of calyx teeth longer
than 2 mm. A white form ✦✧ ✽ is in cultivation, as are various
shades of pink to deep purple. There is also a form with a green leaf,
P. purpurea 'Green leaf' ✦
   Natural hybrid:
      P. × *margaritae* Aparicio & Silvestre
   Lagascalia 14(1):100-101 (1986)

(*P.* × *composita* Pau × *P. purpurea* L.)

Woody sub-shrub 50-80 cm Basal leaves 9-17 × 2.5-3 cm, ovate-lanceolate, truncate or obtuse at base. Floral leaves 5-7 × 1.5-3 cm, ovate-lanceolate, acute. Calyx 18-22 mm, teeth 7-10mm. Bracteoles 15-20 × 0.6-1.3 mm, ovate-lanceolate, densely pilose. Corolla 28-32 mm. Upper lip rose-coloured, lower lip brownish yellow.
Distribution: Spain (Sierra Margarita)

P. purpurea L. **subsp.** *almeriensis* (Pau) Losa & Rivas Goday ex Rivas Mart. ◆⚘

Acta Bot. Malacitana 2:61 (1976)
Colour Plate: VI (iv)
Synonym:
     P. *purpurea* L. var. *almeriensis* Pau
Distribution in the wild: Spain (Almería, Granada).
This subspecies has generally narrower and smaller grey leaves (4–9 × 0.7–3 cm) than P. *purpurea* var. *purpurea* and is found in the Almería region of Spain. Calyx teeth with mucros to 1mm, occasionally 2mm, but much less than in *P. purpurea* var. *purpurea*. Bracteoles shorter.

P. purpurea L. **subsp.** *caballeroi* (Pau) Rivas Mart. ⚘

Acta Bot. Malacitana 2:61 (1976)
Synonym: P. *caballeroi* Pau
Distribution in the wild: Algeria & Morocco.
Described as being from Almería (Spain), and also mentioned in the province of Alicante in Spain. Rivas Goday and Rivas Martínez (1969) began to doubt its presence in Almería. Mateu (1986) has never found living material attributable to this taxon in Spain.
The upper leaf surface is green and almost hairless in this subspecies. Calyx mucros very small and hidden in the indumentum.
Two varieties of *P. caballeroi*, var. *montana* and var. *submontana* have been reported by Pau and Font Quer from Morocco.

Phlomis sieberi Vierh.

Öst. Bot. Zeitschr. 65:231 (1915)
Distribution in the wild: Crete
sieberi is in honour of Franz Wilhelm Sieber (1789-1844) of Prague.

Leaves oblong-ovate-elliptic, 2.5 to 3 times as long as wide, rounded at base, upper surface green, lower canescent. One to two whorls. Calyx 12.5–15 mm, stellate-tomentose; erect teeth 1–2 mm. Corolla yellow, 23 mm.

Phlomis viscosa Poir. ✦

Lam. Encycl. Méth. Bot. 5:271 (1804)

Synonym:

P. *glandulosa* Schenk

Distribution in the wild:

Lebanon, Syria, Palestine and Turkey, in maquis and garigue. On rocky slopes, schistose hillsides and *Quercus* scrub at 300–1440 m.

Flowering in the wild: March–July

viscosa means sticky or viscid. Note however, that some of the clones in cultivation are not as sticky to the touch as for example the perennials P. *samia* and P. *bovei* subsp. *maroccana*.

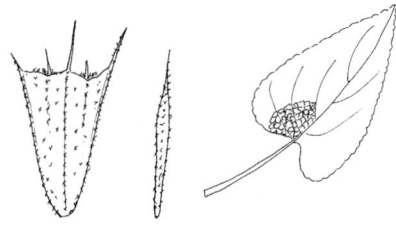

Evergreen shrub 80–150 cm. Leaves glandular-stellate plus simple glandular hairs on upper surface, broadly ovate to oblong-lanceolate, cordate or rounded at base, crenate or serrate at margin, green to somewhat canescent above, canescent below, 4–15 × 1.5–7 cm; petiole to 4 cm. Floral leaves oblong to lanceolate, acute to acuminate, petiolate. One to four distant whorls, 12–20 flowered. Bracteoles numerous, usually in twos or threes, 15–22 × 1.5–4 mm, subulate, hispid-viscid, incurved, ciliate. Calyx shortly pedicellate 18–25 mm, green, stellate-hispid and glandular hairy; teeth subulate, unequal, the two longer ones 4–8 mm. Corolla yellow, about 25–35 mm. Nutlets hairless.

Section *Phlomis* Part 2—Perennials

This listing contains descriptions of all Phlomis perennials known to be in cultivation either privately or in Botanic Gardens—in alphabetical order

Phlomis americana H. Gaud. (invalid?)

This 'species' has appeared in the Index Seminum of a Turkish Botanic Garden, but I can find no trace of a botanical description.

Phlomis angustissima Hub.–Mor.

Bauhinia 1(2):101 (1958)
Distribution in the wild:
   Endemic to Turkey, on dry slopes, limestone and serpentine scree at 1100–2170 m.
Flowering in the wild: June–August
angustissima means having narrow leaves.
Closely related to *P. linearis* Boiss. & Bal. and *P. armeniaca* Willd. It differs from both by its very narrow leaves and the fact that the inflorescence has glandular hairs.

Herbaceous perennial to 60 cm, with glandular and non-glandular hairs. Basal leaves stellate-tomentose, upper surface green or yellowish-green, with ± loose indumentum, lower surface densely whitish stellate-tomentose, glandular hairs sparse or absent, lanceolate to linear-lanceolate, obtuse or subacute, cuneate towards base, entire or indistinctly crenulate at margin, 4–12 × 0.5–1.5 cm; petiole to 8 cm. Upper stem leaves smaller with ± numerous glandular hairs. Floral leaves linear-lanceolate, spreading, to 12 × 0.8 cm, with stellate and numerous glandular hairs. Two to five distant whorls per stem, 4–8 flowered. Bracteoles numerous, linear, 6–12 mm, densely stellate and glandular. Calyx 13–17 mm, yellowish stellate-tomentose, with longer undivided, glandular and non-glandular hairs; teeth ovate-triangular, subulate above, 5 mm. Corolla yellow, 30–35 mm. Nutlets hairless. Hardy to -10°C.

Phlomis anisodonta Boiss.

Diagn. Pl. Or. Ser 1(5):37 (1844)
Distribution in the wild:
   Endemic to Iraq & Iran
anisodonta means unevenly toothed.
An attractive low growing perennial which gradually spreads. Pink and white forms are in cultivation. Because of the white form's colour, there has been some confusion with *P. cancellata,* but *P. anisodonta* has much longer leaf petioles than *P. cancellata* and much shorter calyx teeth (see *P. cancellata*)

Herbaceous perennial 30–40 cm. Basal leaves green above, canescent beneath, ovate to oblong-lanceolate with cuneate to truncate base, lightly crenate at margin, 6–12 × 2–3 cm; petiole to 12 cm. Three to four distant whorls per flower stem, many flowered. Bracteoles subulate, hispid, ciliate. Calyx 12–14 mm, teeth unequal, short, ovate-triangular 2–3 mm, hispid ciliate. Corolla rose-coloured.

Phlomis armeniaca Willd. ✦✧

Sp. Pl. 3:119 (1800)
Illust.: Sweet's *British Flower Garden* :364 (1831-8)
Colour Plate: I (i)
Synonyms:
   P. nympharum O. Schwarz
   P. linearis Boiss. & Bal. subsp. *anticragi* P.H.Davis
Distribution in the wild:
   Endemic to Turkey, in pine woods, steppe, dry limestone rocks, schist
   slopes, corn & fallow fields at 800–2350 m.
Flowering in the wild: June–August
armeniaca means of Armenia.

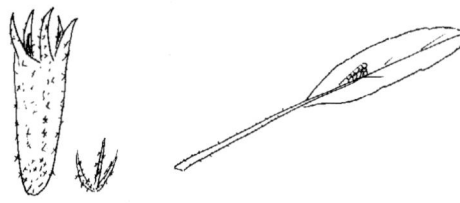

Closely related to *P. linearis* Boiss. & Bal. (see page 49) and *P. angustissima*
Hub.–Mor., (see page 37). Forms a low growing hump of small narrow grey
leaves.

Non-glandular herb 20–60 cm. Woody base. Basal leaves ovate-oblong to lanceolate and
linear-lanceolate, obtuse to sub-acute, cuneate at base, stellate-tomentose, indistinctly
crenulate at margin, 2–10 × 0.8–2 cm; petiole to 7 cm. Floral leaves linear-lanceolate with a
short petiole or stalkless. Two to five distant, upper congested, whorls per stem with 4–10
flowers each. Bracteoles subulate, 3–10 mm, stellate-tomentose. Calyx 13–17 mm, stellate-
tomentose, teeth lanceolate, acuminate, 4–6 mm. Corolla yellow, 25–36 mm. Nutlets hairless.
Hardy to –10°C.

   Natural hybrids:
      P. × bornmuelleri Rech. f.
   Öst. Bot. Zeitschr. 89(4):293,294, t. 8f.×2,295, t 9f.×2,(1940)
   (= *P. armeniaca* Willd. × *P. nissolii* L.)
   Synonym: *P. armeniaca* Willd.var. *subcordata* Bornm.
      P. × rechingeri Hub.–Mor.
   Bauhinia 1(2):103 (1958)
   (= *P. armeniaca* Willd. × *P. carica* Rech. f.)

Phlomis aucheri Decne. ✦

Ann. Sci. Nat. Bot. Sér. 2(2):251 (1834)
Distribution in the wild;
     Endemic to Iran
aucheri is in honour of the French collector P.M.R. Aucher-Eloy (1792-1839)

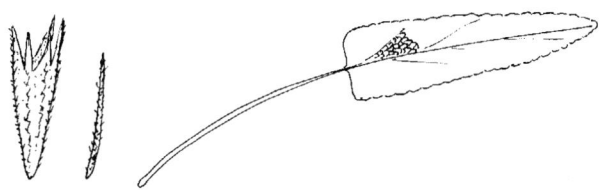

Herbaceous perennial 30–60 cm. Basal leaves narrowly ovate to oblong, rarely lanceolate-linear, cordate or almost truncate at base, 6–12 × 2–3 cm; petiole to 12 cm. Stem leaves shortly petiolate, truncate to cuneate. Three to six distant whorls per simple or branched stem, 5–8 flowered. Bracteoles 13–18 mm, linear, shortly stellate-tomentose. Calyx 16–20 mm narrowly funnel shaped with stellate hairs to 16 mm, teeth unequal. Corolla ±25 mm, yellow.

Natural Hybrid:

P. x stapfiana Rech. f.

Öst. Bot. Zeitschr. 89:293 (1940)

(*P. aucheri* Boiss. × *P. olivieri* Benth.)

Phlomis bovei De Noé

Bull. Soc. Bot. Fr. 2:585 (1855)

bovei is in honour of Nicolas Bové (1812-1841)

Perennial to 1.5 m. Often develops a stout woody base and maintains winter foliage. All parts sticky (dendroid stellate glandular). Basal leaves green, heart-shaped, scalloped margins, 6.5–25 × 4.5–20 cm; petiole 4–18 cm.

P. bovei De Noé subsp. bovei

Bull. Soc. Brot. Fr. 2:585 (1855)

Synonym:

P. samia Desf.

Distribution in the wild:

Algeria, Tunisia

In cultivation but not widely grown. Flower stems have a propensity to flop and so a site with support is called for. It is however very floriferous, with a claret-coloured corolla.

Species to 1. m. Stems with sparse covering of stellate hairs a few of which are glandular. Corolla a dull purple to 3.5 cm in length. Central lobe of lower lip up to 9 mm broad and lateral lobes curve inwards. The calyx is 16–18 × 5mm and almost tubular. Bracteoles incurved round whorl.

P. bovei De Noé subsp. maroccana Maire

Bull. Soc. Hist. Nat. Afrique N. 19:62 (1928)

Illust.: *Bot. Mag.* t. 9144 (1926)

Colour Plate: VIII (iii)

Synonym:

P. samia subsp. *maroccana* (Maire)

Distribution in the wild: Morocco, Algeria

maroccana means of Morocco, but I have added Algeria to the distribution area because the plant detailed in the *Botanical Magazine* (1926) was grown from seed collected in Algeria and is a perfect match to Maire's later description (1928) for the subspecies *maroccana*. This is a very attractive species with magnificent flower heads of lilac colour. Develops a thick woody base. Like *P. samia* most of the plant is sticky although usually less so than *P. samia*.

I feel that the status of this plant should be raised to the level of species, so different is it from *P. bovei* De Noé. Instead of a tightly packed whorl of near tubular calyces surrounded by tightly incurved bracteoles, we have a whorl of pot-bellied calyces widely divergent, loosely surrounded by divergent bracteoles. The corolla is much bigger than in the species and the upper lip is separated from the lower one by some 5 mm in the subspecies. The central lobe of the lower lip is at least twice as wide as for the species. The colours of the corolla are much paler.

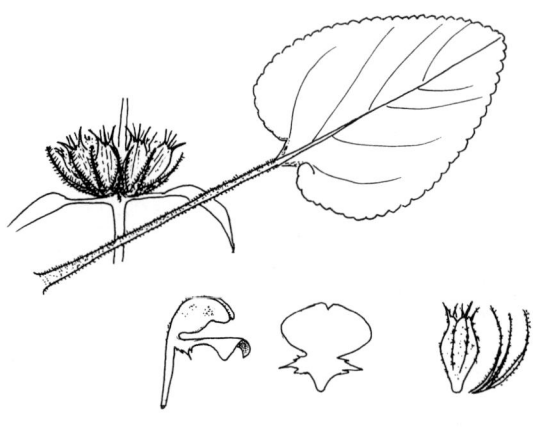

Corolla	Corolla
Upper lip	Lower Lip

To 1.5 m. Stems and lower side of leaves covered in dendroid glandular hairs. Viscose to the touch. The calyx is larger (up to 20 × 7mm), and distinctly pot-bellied rather than tubular as in the species. Corolla much longer, to 4.5cm. lower lip heavily spotted white and purple. Individual corollas within a whorl can vary in colour. Upper lip of corolla well separated (5 mm) from lower lip. Central lower lobe much broader (20 to 28 mm across) than in the species and outer lobes not revolute. Calyces and bracteoles divergent as in species.

Phlomis brachyodon (Boiss.) Zohary

Eig et al. Sched. Fl. Exsicc. Palaest. 23 (1938)
Synonyms:
   P. armeniaca Willd. var. *brachyodon* Boiss.
   P. orientalis Mill. var. *brachyodon* (Boiss.) Boiss.
Distribution in the wild:
   Israel, Jordan, Lebanon & Syria, on rocky steppe.
Flowering in the wild: April–June

brachyodon means with short teeth and refers to the calyx.
Winter leaves are larger and with less hair cover to promote photosynthesis.
P. brachyodon (Boiss.) Zohary **subsp.** ***brachyodon***

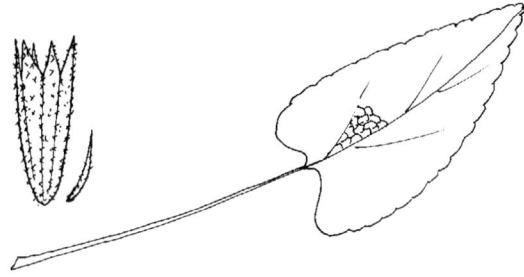

Herbaceous perennial (chamaephyte) 30–50 cm. Leaves obtuse, ovate-oblong to oblong-lanceolate, cordate at base, obscurely crenulate at margin, 5–8 × 2.5–3.5 cm; petiolate. Basal and lower cauline leaves thickly felted (floccose). Floral leaves ovate, not cordate at base. Two to five whorls each 4–8 flowered, lower ones distant. Bracteoles few, lanceolate-subulate, 4–6 mm long. Calyx tubular, densely stellate-hairy, 12–15 mm; teeth triangular-lanceolate to ovate, 3–5 mm. Corolla 25 mm, yellow,

P. brachyodon **subsp.** ***damascena*** (Bornm.) Sam.
Ark. Bot. ser. 2(5) :361 (1960)
Endemic to Lebanon & Syria

Herb 20–35 cm. Lower leaves , cordate to truncate at base, indistinctly crenate at margin, 2.5–5 × 1.3–2 cm; petiolate. Cauline leaves with a shorter petiole. Two to four whorls 4–6 flowered, lower ones distant. Bracteoles few, narrowly subulate to 2 mm (rarely to 8 mm). Calyx tubular 11–14 mm; teeth lanceolate-triangular c. 3 mm. Corolla yellow, 24–26 mm.

Phlomis bruguieri Desf.

Mém. Mus. Hist. Nat. (Paris) 11:9, t.3 (1824)
Distribution in the wild:
   Turkey, Lebanon, Iran, Iraq, Syria, in steppe, fallow fields, gullies at
   edge of plain, at 400–1500 m.
Flowering in the wild: July–August
bruguieri is named after J.C. Bruguière who collected in Persia in 1795.
Not in any collection that I know of, but looks worthy from herbarium
specimens.

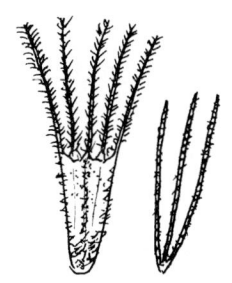

Non-glandular herbaceous perennial to 80 cm. Leaves densely canescent-tomentose, especially beneath, oblong, lanceolate or linear-lanceolate, tapering towards base and acute apex, entire or nearly so at margin, 4–12 × 0.8–2 cm; petiole to 2 cm. Floral leaves shortly petiolate or sessile, linear-lanceolate. Two to five whorls, ± congested, 6 flowered. Bracteoles numerous, subulate, 10–25 mm, densely white-plumose. Calyx 25–35 mm, densely white-tomentose with long spreading hairs, teeth subulate, of unequal length, 15–25 mm. Corolla yellow, 20–25 mm. Nutlets hairless.

Natural hybrids:

P. × *pabotii* Rech. f.

Fl. Iranica 150:316 (1982)
(*P. bruguieri* Desf. × *P. pachyphylla* Rech. f.)
Distribution in the wild: Iran

P. × *praetervisa* Rech. f.

Öst. Bot. Zeitschr. 89:296 (1940)
(*P. bruguieri* Desf. × *P. kurdica* Rech. f.)
Distribution in the wild: Iraq

Phlomis bucharica Regel

Acta. Horti. Petrop. 9,2:579 (1886)
Colour Plate: VIII (ii)
Distribution in the wild:
Afghanistan and former USSR (Central Asia), in high foothills to foothill plains in grassland, in grass and wormwood areas.
Flowering in the wild: June–August
bucharica means of Buchara in Central Asia.

Herbaceous perennial 40–60 cm. Leaves elliptical, lanceolate or ovate-lanceolate, margin mostly entire or sparsely dentate at base. 9–16 × 3–7 cm; petioles of basal leaves 2.5–3.5 cm. Upper leaves similar or narrowly lanceolate, smaller, gradually attenuate to broad petiole, covered with monoradial hairs, densely stellate-tomentose beneath, 4.5 × 8–10 mm. Stems branched, floccose, with four to seven distant whorls of 2–8 flowers. Bracteoles free at base, thickish, attenuate towards apex, appressed to flowers, half to two thirds length of calyx, stellate-hairy. Calyx tubular-campanulate 14–17 mm, floccose-tomentose, erect teeth 2–3 mm, equal. Corolla yellow. Nutlets hairless.

Phlomis cancellata Bunge

Mém. Acad. Scienc. Pétersb. sér. 7, 21: 76 (1873)
Distribution in the wild:
Afghanistan, Iran, Central Asia, Caucasus
Flowering in the wild: May–June
cancellata means cross-barred; latticed and refers to the leaves.
Confused in the trade with the white form of *P. anisodonta* which has longer petioles.

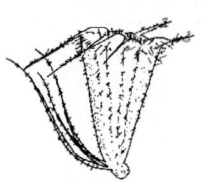

Perennial 20–35 cm. Whole plant stellate-tomentose. Lower leaves oblong, subobtuse, coarsely dentate at margin, upper surface green, rugose, prominently veined, 7–15 cm × 2–4 cm. Upper leaves oblong-lanceolate, coarsely dentate at margin, 7–8 cm × 1.5 cm; petiole to 1–2 cm. Five to seven whorls per stem, lower ones distant, 6–8 flowered on peduncle 3–6 mm. Bracteoles subulate, connate at base in twos or threes, covered in monoradial hairs, 15–17 mm. Calyx tubular 9–11 cm, prominently veined, unequal teeth spreading horizontally, 5 and 10 mm. Corolla white or yellow. Nutlets bearded.

Phlomis capitata Boiss.

Diagn. ser. 2(4):46 (1859)

Distribution in the wild:

Turkey, in. steppe, *Quercus* scrub, rocky slopes, limestone rocks, 540–2,400 m.

Flowering in the wild: June–August

capitata means growing in a dense head.

The species is very variable in the wild. The form with tightly congested whorls is rare and found mainly at higher altitudes.

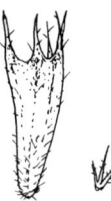

Herbaceous perennial, 10–30 cm. Non-glandular (rarely glandular above). Leaves with short stellate-tomentose indumentum on both sides. Basal leaves ovate to oblong, obtuse, shortly cuneate, truncate or shallowly cordate at base, finely crenate–serrate or entire margin, 2–5 × 1–3 cm; petiole to 7 cm. Cauline leaves similar, shortly petiolate. Floral leaves oblong to lanceolate. Three to six whorls, mostly congested, 3–6 flowered. Bracteoles few, linear-subulate, stellate-tomentose, 3–6 mm. Calyx 10 15 mm, densely stellate-tomentose, with ± numerous longer undivided hairs; teeth lanceolate-acuminate, 3–6 mm. Corolla yellow (rarely with rust brown/ violet upper lip), 20–25 mm. Nutlets hairless , sometimes papillose or hairy at apex.

Phlomis cashmeriana Royle ex Benth.

Hook., *Bot. Misc.* 3:382 (1833)

Illustr.: *Bot. Reg.* 30:22 (1844)

Colour Plate: I (ii)

Synonym:

P. dichroa Rech. f.

Misident.:

P. cashmeriana Royle misapplied in *The New RHS Dictionary*

Distribution in the wild:

Afghanistan, Pakistan, Kashmir

cashmeriana means of Kashmir.

One of the most admired species, this has long narrow leaves, green above and silver beneath. The flowers in bud are spectacular with their speckled appearance and the flower colour is lilac. Quite variable in nature.

Autumn cuttings root OK, but find it difficult to survive the winter in small pots. Therefore propagate in Spring.

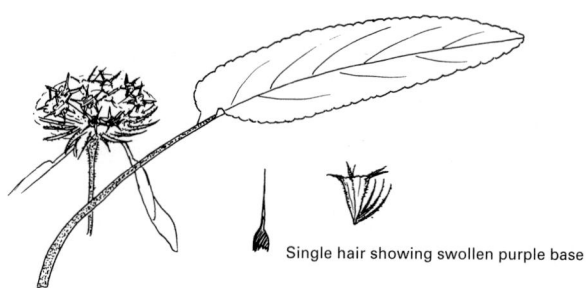

Single hair showing swollen purple base

Perennial to 90 cm. Woody base. Basal leaf laminas long and narrow, dark green above and silvery white beneath, often with an unsymmetrical truncate base, 10–30 × 2–8 cm; petiole 10–15 cm. Plant woody at base and leaves may persist through mild winters. Floral leaves 5–12 × 1.6–4.1 cm; petiole 1–1.5 cm. White woolly stems 40–80 cm, unbranched (occasionally branched), bearing 1–3 whorls with many flowers. Whorls 5– 8 cm across. Bracteoles numerous, ciliate, to 22 × 2 mm, divergent. Calyx c. 18 mm with teeth to 9 mm. Besides the short stellate hairs, the bracteoles, calyx ribs and teeth have long multicellular hairs with bulbous purple bases, which give the whorls a unique speckled appearance especially in the bud stage (hardly visible in dried specimens). Corolla to 20–30 mm, upper lip pale lilac, lower lip purple. Hardy to -10°C.

Phlomis crinita Cav.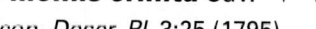

Icon. Descr. Pl. 3:25 (1795)
Colour Plate: I (iii)
Distribution in the wild:
    Spain, Morocco, Algeria and Tunisia
crinita means furnished with long, generally weak hairs.
The species will form a large clump in this country and send out many runners. For the many natural hybrids with *P. lychnitis*—see page 51.

Similar sized sub-shrub (75 cm in height) to *P. nissolii*, but leaves broader, shorter and thicker. Leaves, predominantly basal, 6–18 cm by 2–5 cm, ovate or lanceolate, cordate at base, looking grey because of the woolly stellate-lanate indumentum; petiole to 7 cm and dendroid haired. Upper stem leaves ovate, sparingly petiolate and somewhat connate. Floral leaves usually rhombic (clasping the whorls as in *P. lychnitis*) 2–4 × 1–2.5 cm, sessile. Whorls 6–10 flowered; bracteoles less than a millimetre in width, linear 10–18 mm with long soft hairs up to 3 mm emanating from a dendroid base. Calyx covered with similar long haired dendroid hairs, conspicuous equal teeth to 8 mm, including mucro to 6 mm covered in short stellate hairs. Corolla brownish yellow or yellow. Nutlets hairless. Hardy to −10°C.

    Natural hybrids: —see page 51.
P. crinita subsp. mauritanica (Munby) Murbeck
*Acta. Univ. Lund. Ser.*2, 1(40):66 (1905)
Synonyms:
    P. mauritanica Munby
    P. biloba Desf.

© Jean-Pierre Jolivot

(i) *P. armeniaca* Willd.

© J Mann Taylor

(ii) *P. cashmeriana* Royle ex Benth. [in bud]

© J Mann Taylor

(iii) *P. crinita* hybrid

Plate I

© J Mann Taylor

(i) *P. chrysophylla* Boiss.

© J Mann Taylor

(ii) *P. grandiflora* H.S.Thompson

© J Mann Taylor

(iii) *P. breviflora* Benth.

© J Mann Taylor

(iv) *P. breviflora* Benth.

© J Mann Taylor

(v) *P. lychnitis* L.

© J Mann Taylor

(vi) *P. regelii* M. Popov

Plate II

© J Mann Taylor

(i) *P. leucophracta* 'Silver Janissary'

© J Mann Taylor

(ii) *P. cretica* C. Presl

© J Mann Taylor

(iii) *P. leucophracta* 'Golden Janissary'

© Jean-Pierre Jolivot

(iv) *P. pungens* Willd.

© J Mann Taylor

(v) *P. leucophracta* P.H. Davis & Hub.-Mor.

© Jean-Pierre Jolivot

(vi) *P. nissolii* L.

Plate III

© Jean-Pierre Jolivot

(i) P. 'Edward Bowles' [in winter]

© Jean-Pierre Jolivot

(ii) *P.* 'Edward Bowles'

© Jean-Pierre Jolivot

(iii) *P. cypria* Post var. *occidentalis* Meikle

© J Mann Taylor

(iv) *P. italica* L.

© J Mann Taylor

(v) *P. umbrosa* Turcz. var. *australis* Hemsl.

Plate IV

© J Mann Taylor

(i) *P. tuberosa* L.

© J Mann Taylor

(ii) *P. tuberosa* L.

© J Mann Taylor

(iii) *P. betonicoides* Diels

© J Mann Taylor

(iv) *P. lanata* Willd.

Plate V

(i) *P. samia* L.

(ii) *P. samia* 'Green Glory'

(iii) *P. purpurea* L.

(iv) *P. purpurea* L. ssp. *almeriensis* (Pau)
Losa & Rivas Goday

(v) A white form of *P. purpurea* L.

(vi) A white form of *P. purpurea* L.

Plate VI

© J Mann Taylor

(i) *P. bourgaei* 'Whirling Dervish'

© J Mann Taylor

(ii) *P. russeliana* (Sims) Benth.

© J Mann Taylor

(iii) *P. longifolia* Boiss. & Blanche

© J Mann Taylor

(iv) *P. taurica* Hartwiss ex Bunge

Plate VII

© J Mann Taylor

(i) *P. lunariifolia* Smith

© J Mann Taylor

(ii) *P. bucharica* Regel

© J Mann Taylor

(iii) *P. bovei* De Noé subsp. *maroccana* Maire

© J Mann Taylor

(iv) *P. atropurpurea* Dunn

© J Mann Taylor

(v) *P. atropurpurea* Dunn

Plate VIII

P. crinita var. malacitana Pau

Mem. Mus. Ciencias Nat. Barcelona ser. Bot. 1 (1):64 (1922)

Probably a hybrid with *P. lychnitis* —see page 52

Phlomis fruticetorum Gontsch. ✦

Tr. Tadzh. Bazy. An SSSR, II :186 (1936)

Synonyms:

P. *salicifolia* var. *sewerzowii* Regel

P. *salicifolia* var. *intermedia* Regel

P. *salicifolia* var. *latifolia* Regel

Distribution in the wild:

Former USSR (Central Asia), in high foothills, woodland and scrub belt.

Flowering in the wild: June–July

fruticetorum means partially shrubby.

Only just joined the Collection. It is believed to be quite close to *P. cashmeriana*, but the leaves are more papery and do not seem to have the intense white indumentum on the lower side.

Herbaceous perennial 50–80 cm. Basal leaves oblong-lanceolate or cordate-oblong, subcordate or rarely almost rounded at base, shortly acuminate, denticulate at margin. Stem leaves more or less cordate at base, gradually attenuate towards apex with undulate or denticulate margins; petioles 13–20 cm. Upper surface of floral leaves bright green with scattered stellate and monoradial hairs, lower surface densely stellate-tomentose. Stems simple or branched, many whorls borne on short peduncles. Bracteoles subulate 15–17 mm, up curved. Calyx campanulate 13 mm, white-tomentose and monoradial hairs, teeth subulate, unequal, upper 5–6 mm, lower 10 mm. Corolla pink. Nutlets hairless.

Phlomis herba-venti Aggregate

This aggregate of closely related plants consists of *P. herba-venti*, *P. pungens* and *P. taurica*. Because the aggregate covers such a diverse area, with many of the subspecies confined to a smaller area, they have not been properly correlated and their botanical descriptions listed below overlap, and need further revision.

Phlomis herba-venti L. ✦✧

Sp. Pl. :586 (1753)

Illust.: *Bot. Mag.* t. 2449 (1824)

      Sweet, *British Flower Garden* :74 (1831-8)

herba-venti means herb of the wind, because of the way entire plants sever at the base in the autumn and then behave as tumble weeds. *P. taurica* behaves similarly.

The only European habitat known to Linnaeus was the Narbonne area of south France. [The plant was known as *P. narbonensis* by Joseph Pitton de Tournefort (1656-1708)]. Widely grown since the time of Gerard and grown at the Chelsea Physic Garden by Philip Miller (1768). Susceptible to temperatures of more than 10° (C) frost.

Phlomis herba-venti L. subsp. *herba-venti* ✦✧

Distribution in the wild:

France, Spain, Portugal, Sicily, Italy

Flowering: July to September in the wild

Herbaceous perennial 30–75 cm in height. Stems much branched. Leaves lanceolate or ovate, serrate at margin with a cordate or truncate base; a good mid-green and thin, 6–18 × 3–6 cm; petiolate. Floral leaves ovate or lanceolate, entire, crenate or serrate margins, 4–10 × 1– 3 cm; shortly petiolate or sessile. Two to five verticillasters per stem, some 2 –5 cm apart. 6 –17 flowered. Calyces 8 –15 mm long with teeth to 5 mm. Bracteoles linear 10–20 mm. Corolla purple 15 – 20 (–25) mm. Hardy to –10°C.

Phlomis herba-venti L. **subsp.** ***kopetdaghensis*** (Knorr.) Rech. f.
Flora Iranica 150:309 (1982)
Synonym:
   P. kopetdaghensis Knorr.
Distribution in the wild:
   Iran, former USSR

Bracteoles connate in twos, 12–15 mm, covered in monoradial hairs (the long ray 5 times the length of the others). Calyx teeth spreading, erect, longest 4–5 mm.

Phlomis herba-venti L. **subsp.** ***lenkoranica*** (Knorr.) Rech. f.
Flora Iranica 150:309 (1982)
Synonym:
   P. lenkoranica Knorr.
Distribution in the wild:
   Iran, former USSR

Bracteoles connate in threes 7–9 mm long, covered with monoradial hairs (the elongated ray 10–15 times the length of the others). Calyx teeth spreading horizontally, 3 and 8 mm long.

Phlomis pungens Willd.

Sp. Pl. 3:21 (1800)
Illust.: Sweet, *British Flower Garden* 1:33 (1823-25)
Colour Plate: III (iv)
Synonyms:
   P. herba-venti L. var. *tomentosa* Boiss.
   P. herba-venti L. var. *pungens* (Willd.) Schmalh.
   P. herba-venti L. subsp. *pungens* (Willd.) Maire ex De Fillipps
Distribution in the wild:
   Spain, Morocco, Algeria, Tunisia, Israel & Jordan, Lebanon & Syria,
   Turkey, Albania, Bulgaria, Greece, former Yugoslavia, Iran, Iraq ,
   Crimea and former USSR, in steppe, pastures, fallow fields,
   roadsides, dry stony slopes, *Pinus* forest, low foothills, gullies, dry
   steppes and scrub at 250–2400 m.
Flowering in the wild: May–August
pungens means terminating in a sharp point and applies to the bracteoles.
P. pungens is similar to *P. herba-venti,* but differs in the following respects:
   1) the presence of hirsute hairs on the stems of *P. herba-venti* (easily
visible to the naked eye), and their absence in the case of *P. pungens.* (has a
fine stellate-tomentose indumentum)
   2) the presence, in *P. herba-venti*, of a stellate hair type which has single
hairs up to 3 mm long on the lower surfaces of leaves, bracts, bracteoles and

outer calyx surface. These are absent in *P. pungens*.

Non-glandular herbaceous perennial to 70 cm. Stellate-tomentose hairs to 0.1 mm, ± a secondary indumentum of undivided, articulate 1–3 mm hairs. Cauline leaves lanceolate to ovate-lanceolate, cuneate at base, denticulate or serrate, rarely entire at margin, 5–13 × 1–6 cm; petiole to 10 cm. Floral leaves sessile to shortly petiolate, lanceolate or oblong, acute or acuminate. Two to seven whorls 2–6 (–15) flowered. Bracteoles subulate, 11–17 mm, stellate-tomentose, ± undivided hispid hairs. Calyx 8-15 mm, stellate-tomentose, ± undivided hispid hairs. Corolla purple or pink, 15–20 (–25) mm. Nutlets hairless.

P. pungens Willd. **var. *pungens***

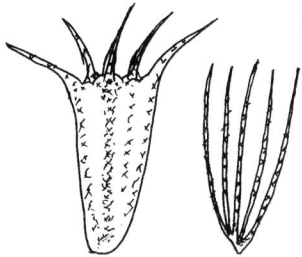

Stellate hairs to 0.1 mm; longer undivided hairs absent or scattered. Calyx teeth half as long as tube or longer. Hardy to –15°C.

P. pungens Willd. **var. *hirta*** Velen.
Fl. Bulg. Suppl. 232 (1898)

Stellate hairs to 0.1 mm; undivided, articulate hairs present, 1–3 mm. Bracteoles and calyx with ± numerous undivided hairs to 2 mm.

P. pungens Willd. **var. *hispida*** Hub.–Mor.
Bauhinia 1(2):118 (1958)

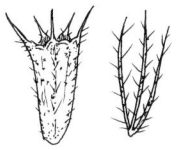

Leaves stellate-tomentose above; stem with dense indumentum of undivided hairs, especially near nodes.

P. pungens Willd. **var. *laxiflora*** Velen.
Sitz-Ber. Böhm. Ges. Wiss. 1887:460 (1888)

Stellate hairs to 0.1 mm; longer undivided hairs absent or scattered. Calyx teeth one third to one fifth as long as tube.

P. pungens Willd. **var. *seticalycina*** (Nábelek) Hub.–Mor.
Bauhinia 1(2):119 (1958)
Synonym:
    P. seticalycina Nábelek

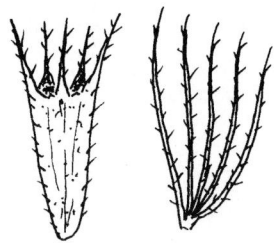

Bracteoles and calyx with very dense indumentum of undivided hairs to 3 mm. Leaves almost hairless above; stem with sparse undivided hairs.

Phlomis taurica Hartwiss ex Bunge
Mém. Acad. Sc. Pétersb. 7 sér. 21, 1:77 (1873)
Colour Plate: VII (iv)
Synonym:
    P. herba-venti f. *euxina* Wassiljevsky
Distribution in the wild:
    Crimea and Caucasus, on stony slopes and scrub.
Flowering in the wild: June–August
taurica means Taurian or of Crimea.
This is here considered as part of a *P. herba-venti* aggregate. A very beautiful plant forming a dome of flowers. The flower buds initially look as if they may be white, but open to a rich mauve. As the stems, leaves and flowers dry out, the whole plant severs at the base and behaves as a tumble weed, blowing away as a complete structure in the wind. It would probably therefore be sensible to have some late flowering bulbs planted beneath it. Hardy to –15°C

Herbaceous perennial 40–50 cm. Lower stem leaves ovate-lanceolate, coarsely dentate at base, upper surface scabrous with scattered simple and stellate hairs, lower surface greyish-green with copious stellate and few monoradial hairs, 11–12 × 4–5 cm; petioles to 4 cm; upper stem leaves oblong-lanceolate, somewhat crenate at margin, 6–7 × 2–2.5 cm, petioles 1–2 mm. Three to four distant whorls per candelabra branched stem, 8–12 flowered. Bracteoles adpressed to calyx, 12– 15 mm, covered in monoradial hairs (elongated ray 10–15 times as long as others). Calyx obconical, covered in monoradial hairs, the veins with simple multi-articulate tubercle-based hairs; calyx teeth subulate, 6–7 mm, erect. Corolla pink. Nutlets white bearded at apex.

Phlomis kurdica Rech. f.
Öst. Bot. Zeitschr. 89:274 (1940)
Synonyms:
    P. orientalis Mill. var. *cordifolia* Nábelek
    P. lanata var. *biflora* Halácsy

Distribution in the wild: Turkey, Israel, Lebanon, Syria, Iran and Iraq, in steppe, corn and fallow fields at 340–2200 m.

Flowering in the wild: May–August

kurdica means 'of the Kurdish lands'.

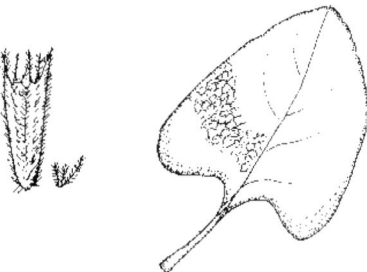

Herbaceous perennial 30–60(–70) cm. Herbaceous perennial 30–60 cm. Leaves stellate-lanate-tomentose (often whitish, especially below) oblong-ovate, obtuse, profoundly cordate at base, crenate margins, 5–12 × 3–8 cm; petiole to 11 cm. Floral leaves cordate-triangular to lanceolate, shortly petiolate. Three to seven whorls (lower ones distant) per stem, 8–10 flowered, stems simple or pyramidally branched. Bracteoles few, sometimes none, lanceolate, 1.5–5 mm, densely floccose. Calyx 15–18mm, densely stellate-tomentose with long spreading hairs, teeth nearly equal, subulate, 4–6 mm. Corolla yellow, 30–35 mm. Upper lip nearly as long as lower; width of lower lip 20 mm or more. Nutlets hairless.

Natural hybrids:

P. x melitensense Hub.–Mor.

Bauhinia 1(2):111 (1958)

(= *P. kurdica* Rech. f. × *P. oppositiflora* Boiss. & Hausskn.)

P. x praetervisa Rech. f.

Öst. Bot. Zeitschr. 89:296 (1940)

(*P. bruguieri* Desf. × *P. kurdica* Rech. f.)

P. x tunceliensis Hub.–Mor.

Bauhinia 6(30:374, (1979)

(= *P. kurdica* Rech. f. × *P. linearis* Boiss. & Bal.)

Phlomis linearis Boiss. & Bal. ✦

Diagn. ser. 2(4):46 (1859)

Distribution in the wild:

Endemic to Turkey, in steppe, rocky igneous slopes, volcanic rubble at 1350–2400 m.

Flowering in the wild: June–August

linearis means narrow with both sides parallel and refers to the leaves.

Closely related to *P. armeniaca* Willd. and *P. angustissima* Hub.–Mor. Distinguished from them by its calyces which are densely hairy and not constricted above.

Perennial 10–30(–40) cm, ± non-glandular. Woody base. Basal leaves greenish above, greyish tomentose below, basal leaves linear-lanceolate, acutish, 2–12 × 0.7–2 cm, tapering into a petiole to 5 cm. Stem leaves smaller, shortly petioled. Floral leaves linear to linear-lanceolate. One to four whorls per stem, 4–10 flowered. Bracteoles linear-subulate, 8–12 mm, stellate-tomentose. Calyx 12–14 mm, densely yellow tomentose ; teeth ovate-triangular, acuminate, 3–7 mm. Corolla yellow, 30–35 mm. Nutlets hairy at apex.

P. linearis var. *plumosa* Boiss. ◆

This variety has an extremely feathery calyx covering. Hardy to –15°C
Natural hybrids:

P. × *kalanensis* Hub.–Mor.
Bauhinia 6 (30: 374, (1979)
(= *P. linearis* Boiss. & Bal. × *P. oppositiflora* Boiss. & Haussk.)
Differs from *P. linearis* with whorls 2–6 flowered, wide teeth and bracteoles 2–8 mm long.
Differs from *P. oppositiflora* with whorls of 2–8 flowered, calyx not tomentose, bracteoles developed.

P. × *tunceliensis* Hub.–Mor.
Bauhinia 6 (30:374, (1979)
(= *P. kurdica* Rech. f. × *P. linearis* Boiss. & Bal.)
Differs from *P. kurdica* by having narrower lanceolate leaves, truncate at base; bracteoles to 10 mm long.
Differs from *P. linearis* by having wider leaves, base truncate to subcordate.

Phlomis lychnitis L. ✧❀

Sp. Pl. 585 (1753)
Illustr.: *Bot. Mag.* t. 999 (1807)
Colour Plate: II (v)
Distribution in the wild:
    France, Spain & Portugal,
lychnitis is the name of a herb mentioned by Pliny.
In Spain the leaves have been used for lamp-wicks and the common name is Candelera.
This species benefits from protection against the worst of winter's wet with glass or cloche. The long narrow grey leaves are very characteristic, as are the rhombic floral leaves or bracts which tightly clasp the whorls in bud.
Sometimes used to adulterate sage (*Salvia officinalis*)

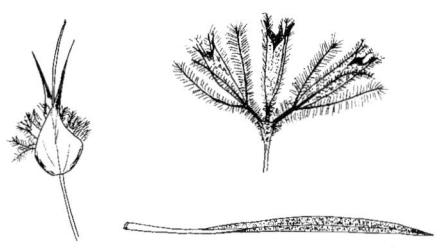

Sub-shrub to 65 cm. Leaves linear, margins entire, upper surface grey/green, stellate hairy and puckered to 15 × 1 cm, arranged in tufts. Leaves taper to base which is flat on upper surface. Flowering stems unbranched, bearing up to six whorls per stem each with up to 10 flowers. Whorls 4–5 cm across. Floral leaves 3–6.5 × 1.8–2.5 cm, upright, rhombic, stalkless, and clasping the calyces in bud. Bracteoles linear 10–20 × 1 mm. Calyx to 20 mm with 5 veins. Calyx and bracteoles covered in long silky multijointed hairs arising from stellate bases. Glandular hairs also present; teeth 2–6 mm. Corolla 20–30 mm, yellow. Nutlets hairless.

One variety (**P. lychnitis var. virens**) has been described based on the greener colour of the upper surface of the leaves and plant's greater size. Mateu's view is that this variety falls within the variability of the species and should be included in the synonymy.

Natural hybrids of *P. crinita* Cav. and *P. lychnitis* L.:
They have leaf-shapes between the two species. The most recent descriptions (Mateu 1986) rely heavily on the hair type present, but in my experience this is not borne out in natural populations and so I rely on the leaf shape as drawn by Pau:

Parent 1:

P. crinita Cav. [See drawing (a)] ✦

Icon. Descr. Pl. 3:25 t. 247 (1795)

Leaves very thick, ovate, cordate at base, 6-18 × 2-5 cm. Calyces with 10 veins. Flower stems to 75 cm.

P. × composita Pau   [See drawing (b)] ✦

Bol. Soc. Arag. Ci. Nat. 17:132 (1918)

Leaves very thick, oblong, somewhat cuneate at base, 3-21 × 0.5-5.5 cm Calyces with 5 or 10 veins. Flower stems to 100cm.

P. × trullenquei Pau [See drawing (c)] ✦

Butll. Inst. Catal. Hist. Nat. 18:161 (1918)

Synonym:

P. × composita Pau nm. *trullenquei* (Pau) Mateu

Leaves thick and velvet like, oblong, decurrent with petiole . 8-20 × 1.5- 2.5 cm. Calyces with 5 veins. Flower stems to 100 cm.

P. × almijarensis Pau [See drawing (d)] ✦✧

Mem. Mus. Ci. Nat. Barcelona 1(1):64 (1922)

Synonym:

P. × *composita* Pau nm. *almijarensis* (Pau) Mateu
Leaves linear or elliptic and thicker than *P. lychnitis*. Length: width
< 7. Calyces with 5 veins. Stems to 75 cm.

Parent 2:

P. lychnitis L. [See drawing (e)]

Sp. Pl. 585 (1753)
Leaves linear 15 × 1 cm. Calyces with 5 veins. Flower stems to 65 cm.
Note: The following variety of *P. crinita* Cav. described by Pau is
probably another hybrid between the above two parents with a
leaf somewhat between Drawing (a) and (b)

P. crinita var. malacitana Pau

Mem. Mus. Ci. Nat. Barcelona 1(1):64 (1922)
Synonym:
    P. × *composita* Pau nm. *malacitana* (Pau) Mateu
Leaves ovate, cordate at base, extra thick indumentum, 9-15
× 2.5-4 cm. Simple hairs dominate. Calyces with 10 veins.
Flower stems to 75 cm.

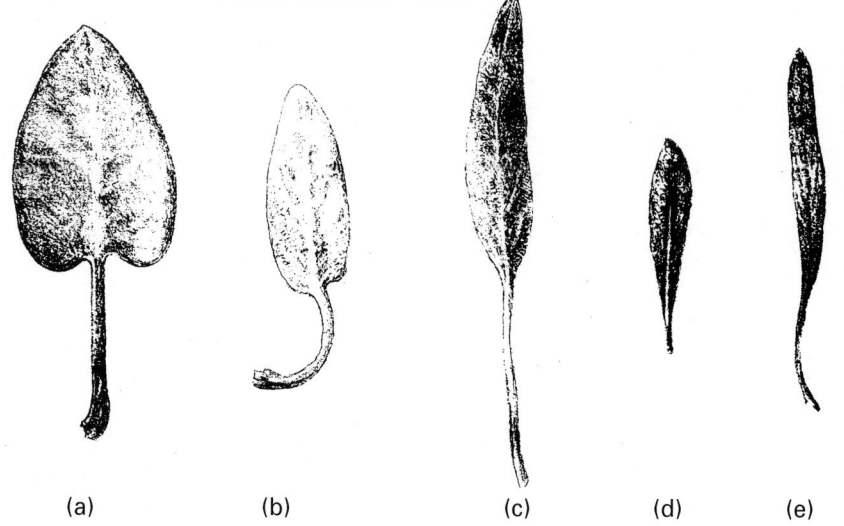

(a)   (b)   (c)   (d)   (e)

Phlomis nissolii L. ✦✧

Sp. Pl. 585 (1753)
Colour Plate: III (vi)
Synonym:
    P. armeniaca Willd. var. *subcordata* Bornm.
Distribution in the wild:
    Turkey, Lebanon and Syria, on limestone scree and rocks, roadsides,
    corn and fallow fields at sea level to 1550 m.
Flowering in the wild: June–August

nissolii is in honour of Guillaume Nissolle (1647-1735).
This has medium thick woolly leaves (rather like *P. crinita*). Winter leaves are
much less woolly and look greenish (the summer wool is a protection against
evaporation). My collection has not flowered in the UK after three years.

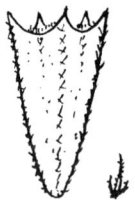

Non-glandular perennial to 100 cm. Woody base. Basal leaves densely white stellate-pannose
on both sides, basal leaves large, oblong to ovate, obtuse, cordate at base, crenulate to
crenate at margin, 6–14 × 3–6 cm; petiole to 12 cm and dendroid haired. Stem leaves smaller,
oblong, truncate or shortly cuneate at base, short petiole. Floral leaves lanceolate. Five to
seven congested whorls per stem, 4–8 flowered. Bracteoles few, subulate, 5–10 mm,
tomentose. Calyx 10–15 mm, densely white stellate-pannose; teeth ovate-triangular,
mucronate, 2 – 4mm. Corolla yellow, 20–30 mm. Nutlets hairless or ±papillose or pilose at
apex. Hardy to –10°C.

Natural hybrid:
P. x bornmuelleri Rech. f.
Öst. Bot. Zeitschr. 89(4):293,294, t.8f.×2,295, t9f.×2,(1940)
(= *P. armeniaca* Willd. × *P. nissolii* L.)

Phlomis olgae Regel

Pl. Nov. Fedsch. 34,2:68 (1881)
Synonym:
   P. tomentosa Regel
Distribution in the wild:
   Former USSR (Endemic to Central Asia), in mountain areas, stony
   slopes, juniper woods and fescue steppes.
Flowering in the wild: May–July
olgae is named after Olga Fedtschenko (1845-1921)

Herbaceous perennial, 30–50 cm. Basal and cauline leaves elliptical or ovate, cordate or
rounded at base, 12–15 × 5–7 cm. Floral leaves compound, entire margin, greyish with stellate
and monoradial hairs, niveous-tomentose below, 7.5 × 3.5 cm. Stems branched in upper
part and stellate-tomentose. Three to five whorls, distant, (2)4–8 flowered. Bracteoles broad
at base, spinous-tipped, free, 18–30 mm long, white with stellate and monoradial hairs.
Calyx campanulate, expanded upward, stellate-tomentose 13–18 mm with teeth 3–5 and 8–
12 mm long, spinescent, horizontally spreading. Corolla purple. Nutlets hairless.

P. olgae var. sarawschanica Regel

The bracts are connate at the base in 2s or 3s.

Phlomis oppositiflora Boiss. & Hausskn. ✦

Fl. Or. 4:784 (1879)
Distribution in the wild:
   Endemic to Turkey, on calcareous slopes and steppe at 910–1500 m.

Flowering in the wild: June–July
oppositiflora refers to the fact that this plant has only two calyces one opposite
the other on the stem. A curiosity rather than a useful garden plant.

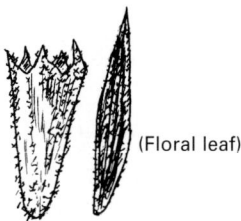

(Floral leaf)

Herbaceous perennial 25–50 cm. Leaves linear-lanceolate to linear, tapering to base,
acutish apex, stellate-tomentose or almost hairless, 4–7 × 0.5–1 cm; petiole to 4 cm. Floral
leaves linear. Four to five (occasionally nine) distant whorls per stem, each with two
opposite flowers. Bracteoles absent (floral leaf shown above). Calyx campanulate, shortly
pedicellate, 10–18 cm, densely lanate, teeth broadly triangular to 3 mm. Corolla 25–35
mm, yellow. Nutlets densely hairy at apex.

Natural hybrid:
P. × kalanensis Hub.–Mor.
Bauhinia 6 (30: 374, (1979)
(= *P. linearis* Boiss. & Bal. × *P. oppositiflora* Boiss. & Hausskn.)
Differs from *P. linearis* with whorls 2–6 flowered, wide teeth and
bracteoles 2–8 mm long.
Differs from *P. oppositiflora* with whorls of 2–8 flowered, calyx not
tomentose, bracteoles developed.
P. × melitensense Hub.–Mor.
Bauhinia 1(2):111 (1958)
(= *P. kurdica* Rech. f. × *P. oppositiflora* Boiss. & Hausskn.)

Phlomis persica Boiss.

Diagn. ser. 1, 5:37 (1844)
Synonym:
    P. cymifera Boiss.
Distribution in the wild:
    Endemic to Iran
persica means of Persia (Iran).

Multi-stemmed 30–40 cm, stellate-tomentose. Lower leaves linear, acute apex, 10 × 2 cm;
petiole 1–2 cm. Numerous whorls, lower ones with a peduncle. Bracteoles subulate, stellate-
tomentose. Calyx tubular, 10 mm; teeth shortly triangular, unequal, ending in a mucro 1–3
mm. Corolla rose coloured.

Phlomis platystegia Post

Bull. Herb. Boissier 1:407 (1893)
Plate:
Synonym:
    P. imbricata Boiss.

Distribution in the wild:
    Israel, on rocky slopes in deserts, usually with *Origanum dayi*.
Flowering in the wild: March–June
platystegia means with broad bracteoles.

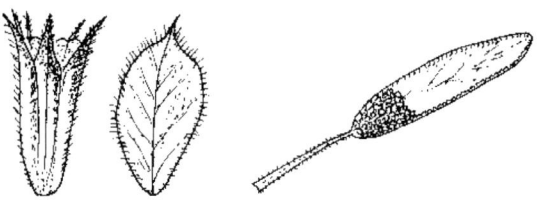

This has leaves somewhat reminiscent of *P. leucophracta* with its white edges, but are much narrower and grey speckled. Very attractive foliage, which may survive the winter (making it look like a shrub–even more so as it often branches from a single stem) or may be cut back to 20cm in winter. Winter leaves are larger and hair cover thinner to promote photosynthesis. Will not survive frost without conditioning for two winters.

Perennial (chamaephyte) 50–100 cm. Stems woolly floccose (stalked stellate). Leaves oblong to ovate-oblong, crenate at margin, 3–10 × 2–4 cm. Cauline leaves petiolate, obtuse, cuneate to cordate at base, green above, canescent below. Floral leaves sessile. Three or more distant whorls, many flowered, per stem. Bracteoles numerous, flat, ovate to oblanceolate, acute, 3–10 mm broad, and nearly as long as calyx. Calyx tubular-campanulate, 12–15 mm, golden stellate hairy, partly glabrescent at maturity. Corolla golden yellow 24–30 mm.

Phlomis pungens Willd.—see *Phlomis herba-venti* aggregate, page 46.

Phlomis regelii M. Popov
Bull. Univ. As. Centr. Suppl. 12:23 (1920)
Synonym:
    P. olgae var. *subcrenata* Regel
Colour Plate: II (vi)
Distribution in the wild:
    Former USSR, in foothills, on gentle slopes, on loess soils.
Flowering in the wild: May–July
regelii commemorates Eduard August von Regel (1815-1892).
Needs ground moisture in the spring to initiate growth. It is slow in growth.

Herbaceous perennial 30–50 cm. Lower leaves lanceolate, 10–14 × 3.5–6 cm, petiole 4–5 cm; floral leaves 5–6 × 2.5–3 cm, upper side scattered stellate and monoradial hairs, lower side stellate-hairy. Four to five whorls (on short peduncle) per stem which is covered in stellate-tomentose hairs. Bracteoles mostly patent or reclinate, 20–25 mm, thick at base, spinescent, stellate-hairy. Calyx tubular-campanulate, 10–11 mm; teeth 3–5 and 8–10 mm long and stellate hairy. Corolla lilac-pink, 15–17 mm. Nutlets virtually hairless.

Phlomis rigida Labill.   ✦❀

Icon. Pl. Syr. 3:15, t.10 (1809)
Illust.: *Icones Plantarum, Syriae Rariorum* t. 10 (1791)
Synonym:
  P. shepardi Post
Distribution in the wild:
  Lebanon, Turkey, Syria, Iran and Iraq, in steppe, *Pinus* woods, *Quercus*
  scrub, and stony gullies, at 300–2000 m.
Flowering in the wild: June-September
rigida means stiff or rigid and refers to the flower stem.
This name which refers to a large rose flowered perennial has been mis-
applied by one large UK nursery which obtained a plant through Cambridge
Botanic Gardens which was in fact the small growing yellow flowered *Stachys
citrina*.

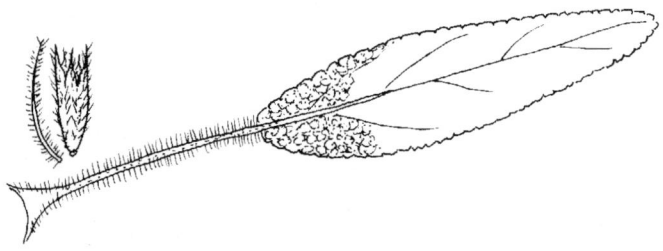

Herbaceous perennial 40–125 cm. Leaves pale green, wrinkled, oblong to lanceolate, obtuse,
shortly cuneate, rounded or truncate at base, crenulate to crenate at margin, 5–30 × 2–10
cm; petiole to 10 cm. Floral leaves lanceolate, petiolate. Flowering stems 50 to 125 cm,
bearing three to five distant whorls of many flowers. Bracteoles numerous, subulate, curved
upwards, 20–25 mm, with dense rigid hairs. Calyx 15–23 mm, stellate-tomentose with a
secondary indumentum of rigid hairs to 10mm, teeth ovate-truncate at base then subulate,
6–10 mm. Corolla pink, 40–50 mm. Nutlets with short hairs at apex. Hardy to –10°C.

Phlomis russeliana (Sims) Benth.   ✦✧❀

Lab. Gen. Sp. 629 (1834)
Illust.: *Bot. Mag.* t. 2542 (1825)
Colour Plate: VII (ii)
Synonyms:
  P. lunariifolia Smith var. *russeliana* (Sims) Benth.
  P. superba C Koch
Misidents.:
  P. samia misapplied
  P. viscosa misapplied
Distribution in the wild:
  Endemic to Turkey, in coniferous and deciduous woods and clearings,
  Corylus scrub. At 300–1700 m.
Flowering in the wild: May–September
russeliana is in honour of Dr Alexander Russell (c. 1715 - 1839), author of a
Natural History of Aleppo.

One of the most popular *Phlomis* in the UK, this plant has very large heart-shaped green leaves on long petioles. The basal leaves survive the winter. The erect unbranched flower stems have bicoloured flowers (cream and yellow) and are decorative in seed right through till the spring. Its one fault(?) is its ability to spread fairly quickly.

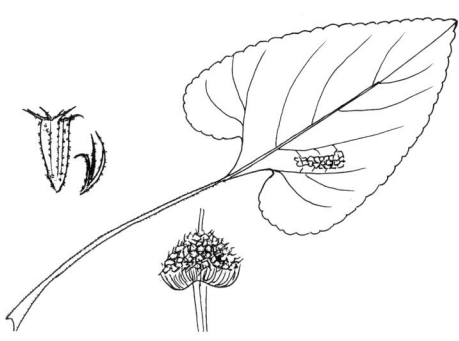

Non-glandular perennial to 1½ m; basal leaves persist throughout winter. Leaves short stalked stellate above and below, green above, whitish beneath, basal leaves broadly ovate-cordate, obtuse with coarse crenate margin, 6–20 × 6–12 cm; petiole to 28 cm with dendroid hairs. Floral leaves 6–21 × 2–11 cm, stalkless. Unbranched flowering stems bearing up to 5 whorls each 12–20 flowered. Whorls 5–8 cm across. Numerous bracteoles 10–20 × 1–2 mm, incurved around calyx. Calyx 20–25 mm with teeth 2–5 mm, splayed outwards. Calyx and bracteoles covered in dense stellate hairs. Corolla 30–35 mm, upper lip of corolla cream, lower lip golden yellow. Central lobe of lower lip to 1 cm across. Nutlets hairless. Hardy to –15°C

Phlomis samia L.

Sp. Pl. 585 (1753)
Illust.: *Bot. Mag.* t. 1891 (1818)
        Andrew's *Botanist's Repository* Pl. 584 (1797-1812)
Colour Plate: VI (i), (ii)
Synonym:
    P. samia var. *graeca* Bornm.
Distribution in the wild:
    Former Yugoslavia, Greece, Turkey, in *Pinus* and *Cedrus* forest,
    metamorphic soils, volcanic slopes, 400–1750 m.
Flowering in the wild: June–August
samia was applied by Tournefort for a collection from the island of Samos. This has moderate-sized green heart shaped leaves and almost all parts of the plant [leaves (lower surface), stem and whorls] are sticky to the touch. The calyces of the Greek plants are often coloured purple especially in the upper parts. *P. samia* differs from *P. bovei* subsp. *maroccana* by not having its woody sub-structure and by the corolla not having the large gap between upper and lower lip or the very wide lower lip of *P. bovei* subsp. *maroccana*. Difficult to split. Propagate from basal cuttings in spring.

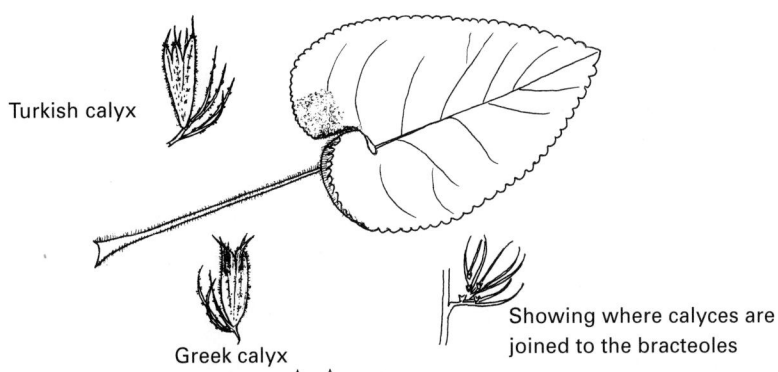

Turkish calyx

Greek calyx

Showing where calyces are joined to the bracteoles

***P. samia* 'Green Cap'** ✦✧ is a selection from Turkey where the upper lip is green coloured.

***P. samia* 'Green Glory'** ✦❀

This is a selection from Turkey where both the upper and lower lips are of a lime green colour. Colour Plate: VI (ii)

Glandular perennial to 1½ m which can maintain some leaves through the winter. Most parts glandular hairy and most parts sticky to the touch (except upper surface of leaves). Basal leaves green with stellate hairs, lanceolate-ovate to broadly ovate, cordate or sagittate at base, crenate or serrate at margin, 8–27 × 5–15 cm; petiole to 30 cm covered in simple glandular hairs. Floral leaves ovate or lanceolate, acuminate, 4–8.5 × 2.5–6.5 cm, shortly petiolate. Flower stems much branched like a candelabra, each with 3–7 whorls, 6–22 flowered. Whorls 6–7 cm across. Bracteoles 20–26 mm long, 1 mm wide, often connected in a pair and a triplet which branch above their base like a tree. Calyx tubular 18–25 mm long and 5 mm wide. Calyx teeth subulate, 6–12 mm. Bracteoles, calyx and teeth covered in glandular stellate hairs. Bracteoles and calyces divergent. Corolla to 35 mm, upper lip of corolla green to purple, lower lip claret coloured to green. Lower lip outer lobe margins rolled downwards and inwards. Central lobe of lower lip to 16 mm wide. Nutlets hairless or topped with short glandular hairs . Hardy to −15°C.

Phlomis syriaca Boiss.

Diagn. ser. 1(12):89 (1853)
Synonym:
    P. nissolii L. var. *leptorrhacos* Boiss.
Distribution in the wild:
    Israel, Lebanon, Turkey. Also Jordan, Syria, on dry hills, steppe. 130–1580 m.
Flowering in the wild: May–July
syriaca indicates its origin in Syria.

Non glandular herbaceous perennial (with persistent woody base) 50–100 cm. . Basal leaves cordate or truncate at base, very thick when young. Cauline leaves oblong to lanceolate, cuneate or cordate at base, crenate at margin, 4–10(–14) × 1.5–2.5(–6) cm. Few to 7 distant whorls, per stem, simple or pyramidally branched, 4–6 flowered. Bracteoles few, subulate, 2–6 mm. Calyx 10–15 mm, tubular or tubular-campanulate, 10 nerves.; teeth ovate, acute, unequal, often with a short mucro. Corolla yellow, 20–30 cm. Upper lip shorter than lower.

Phlomis taurica Hartwiss ex Bunge—see under *Phlomis herba-venti* aggregate, page 48.

Phlomis thapsoides Bunge

Mém. Sav. Etr. Pétersb. 7:440 (1851)
Synonym:
  P. thapsoides Bunge var. *brevisubulata* M. Popov
Distribution in the wild:
  Turkestan, in foothills, grass and wormwood steppes.
Flowering in the wild: June–August
thapsoides means resembling the mullein *Verbascum thapsus*.

Herbaceous perennial 30–60 cm. Stem leaves cordate-ovate, coarsely dentate at margin, acuminate, upper side light green, strigose, scabrous, sparsely hairy, lower side white with closely adpressed monoradial hairs, 7–12 × 3–6 cm; petioles 1.5–4 cm. Floral leaves compound, entire margin, 5–7 × 3–3.5 cm, the uppermost 2.5–3 × 1.5 cm; petioles 0.5–0.8 cm. Five to six whorls per strongly branched stout stem, 2–8 flowered. Bracteoles connate at base in twos, rarely in threes, thick, acuminate, 14–18 mm. Calyx 10–16 mm, densely covered in hairs, teeth equal, acuminate. Corolla lilac. Nutlets hairless.

Section *Phlomoides*

This listing contains descriptions of all Phlomoides known to be in cultivation either privately or in Botanic Gardens—in alphabetical order.

Phlomis agraria Bunge

Ledeb., *Fl. Alt.* 2:411 (1830)
Synonym:
  Phlomoides agraria (Bunge) Adylov, Kamelin & Makhmedov
Distribution in the wild:
  China, former USSR (Kazakhstan, Mongolia, Siberia), in steppe, and
  forest steppe, slopes and in plains, needle grass and fescue steppes
  and scrub.
Flowering in the wild: June–July
agraria means growing in the fields.
Differs from *P. tuberosa* by having very small floral leaves, stellate hairs on the under side of leaves and glandular hairs on calyx around teeth. Many of the plants in cultivation satisfy the hair requirement, but have long floral leaves and may be hybrids with *P. tuberosa*.

> Herbaceous perennial 40–60 cm. Ropelike roots with some tubers. Basal leaves triangular-cordate, upper surface green with simple hairs, lower surface stellate plus simple hairs, obtuse rounded teeth at margin, 8–10 × 4–6 cm; petioles 5–8 cm. Lower stem leaves, 5.5 × 3–3.5 cm; petioles 2–2.5 cm. Upper stem leaves similar but smaller, sessile. Floral leaves 1.5 × 0.6–0.8 cm, ovate, acuminate. All leaves have simple hairs above and stellate plus simple hairs beneath. Many whorls on simple or branched stems, 10–12 flowered. Flowering parts glandular hairy. Bracteoles linear-subulate, spinescent, 8–9 mm, covered with multi-articulate spreading hairs. Calyx tubular-campanulate, 10–12 mm; teeth semiorbicular or rounded subulate-pointed; covered with small stellate hairs and long and short articulate hairs on the veins, with some glandular hairs. Corolla pink, rarely white. Nutlets hairy at apex.

P. agraria Bunge **f. alba** Trautv.
A white coloured form.

Phlomis alpina Pall.

Acta Acad. Petrop. 2:265 (1779)
Synonym:
  Phlomoides alpina (Pall.) Adylov, Kamelin & Makhmedov
Distribution in the wild:
  China, former USSR (Central Asia, Siberia), in alpine and forest
  zones, in meadows.
Flowering in the wild: June–August
alpina means of alpine regions.
Much shorter than *P. tuberosa* and without its large tubers. Flowering stems are green and always unbranched.

> Herbaceous perennial 20–50 cm. string-like roots. Basal leaves ovate, cordate at base, 13–15 × 10 cm; petioles longer than lamina. Stem leaves ovate-oblong to linear-lanceolate, 10 × 3-5 cm. Upper stem leaves short petioled. Lower floral leaves ovate-oblong or oblong-lanceolate, with rounded teeth at margin, 7–11 × 2–4 cm. Upper floral leaves linear-lanceolate,

with obtuse teeth or entire at margin, covered on both sides with scattered 1-jointed hairs. Many whorls per unbranched stem which is almost hairless at bottom and covered with soft retrorse or stellate hairs above. Bracteoles slightly incurved, narrowly linear, 9–11 mm, clothed with long spreading multi-articulate hairs. Calyx campanulate; teeth rounded-ovate terminating in a subulate point, 2–3 mm. Calyx with small fine hairs with larger multi-jointed hairs. Corolla pink. Nutlets hairy at apex.

Phlomis atropurpurea Dunn

Notes Roy. Bot. Gard. Edinb. 8:169 (1913)
Colour Plates: VIII (iv), (v)
Synonym:
   Phlomoides atropurpurea (Dunn) Kamelin & Makhmedov
Distribution in the wild:
   China (Yunnan), in marshy meadows and open pasture at 2800-3900 m. *atropurpurea* means deep purple.
Flowering in the wild: June–August
The form in commerce was found in boggy acid soil. It requires a continuously damp soil through the growing season to do well here, but soil type doesn't seem to be important. It is a dainty plant which also does well in a pot if kept moist. Hardy to –15°C.

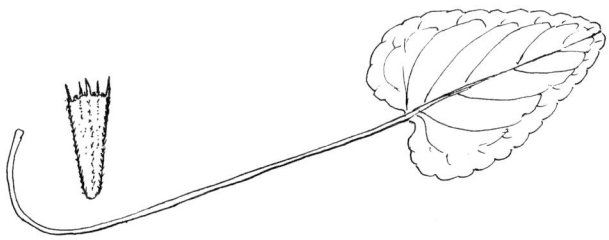

Herbaceous perennial to 20–30 (–60) cm. Thickened roots. Basal leaf laminas glossy green, hairless, ovate, cordate at base, crenate at margin, 2.5–11 × 1.5–8 cm; petiole 2–28 cm. Floral leaves (1.5– 4 × 0.4-2.2 cm) serrate or subentire; short petiole or sessile. Flowering stems (20–60cm) bearing 1–4 whorls each with many flowers. Bracteoles linear-subulate 3–10 × 2 mm, sparsely ciliate, almost hairless. Calyx tubular to campanulate, to 12 mm, teeth 2–3 mm. Corolla 17–20 mm, dull purple. Central lobe of lower lip 7 × 9 mm. Nutlets hairless.

P. *atropurpurea* Dunn **f. *pallidor*** C.Y. Wu
A pale coloured form.
P. *atropurpurea* Dunn **f. *pilosa*** C.Y. Wu
A form covered with long soft hairs.

Phlomis betonicoides Diels ✦✧❀

Notes Roy. Bot. Gard. Edinb. 5:241 (1912)
Colour Plate: V (iii)
Synonym:
   Phlomoides betonicoides (Diels) Kamelin & Makhmedov
Distribution in the wild:
   China (Sichuan, Xizang, Yunnan), on grassy slopes, forested

grasslands, forests, 2700–3000 m.
Flowering in the wild: June–August
betonicoides means with betony like leaves.
If the spring is very dry, this plant can be very late to break through the soil
(June/July) so care should be taken not to hoe it away. A wet spring (or
watering) will bring it through much earlier.

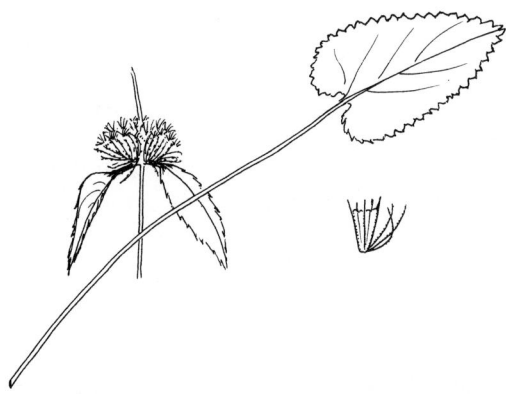

Herbaceous perennial 30–80 cm. Roots like a string of beads. Basal leaves green, narrowly
ovate to triangular, rounded to cordate at base, crenate to dentate at margin, 7–16 × 5–9 cm;
petioles 3–16 cm. Stem leaves 5–9 × 2–4.5 cm. Floral leaves 2.5–8 × 1–3.5 cm, toothed and
sessile. Flowering stems unbranched, 30–80 cm, bearing up to 8 whorls of many flowers.
Bracteoles dark purple, linear subulate to 13 mm, ciliate. Calyx tubular to campanulate, to 13
mm; teeth, spreading, 2–4 mm. Corolla purple, to 18 mm. Nutlets hairless. Hardy to –15°C
P. betonicoides Diels **f. alba** C.Y. Wu
A white form.

Phlomis bracteosa Royle ex Benth.

Hook., *Bot. Misc.* 3:383 (1833)
Synonyms:
P. lamiifolia Royle ex Benth.
P. latifolia Royle ex Benth.
P. simplex Royle ex Benth.
P. cordata Royle ex Benth.
Distribution in the wild:
Afghanistan to SW China.
bracteosa means with numerous or conspicuous bracts (bracteoles)

Herbaceous perennial 20–100 cm, but more commonly to 50cm. Leaves ovate, deeply cordate
to truncate, ± crenate at margin, aromatic, 5–10 × 4–7 cm, petiole 3–8 cm (less for upper
leaves). Three to five whorls 14–20 flowered Bracteoles numerous, linear or lanceolate,
acuminate to spinulose, fimbriate with long simple hairs only or with irregularly rayed
branched hairs. Calyx 10–12 mm indumentum as for bracteoles, sometimes glabrescent;
teeth 3–4 mm, subulate to spinulose, ciliate. Corolla 15–20 mm, rose to purple to violet.

Phlomis breviflora Benth. ✦✧❀

Wallich, *Pl. Asia Rar.* 1:62 (1830)
Colour Plates: II (iii), (iv)
Synonyms:
   Phlomoides breviflora (Benth.) Kamelin & Makhmedov
   P. parviflora Benth.
   P. setigera var. *filiformis* Hook. f.
Distribution in the wild:
   Himalayas (Nepal to Bhutan), India, N. Burma, at 3,000–3,600 m.
breviflora means with a short flower.
Does well in a moist soil. Attractive with its whitish flowers and purplish
calyces.

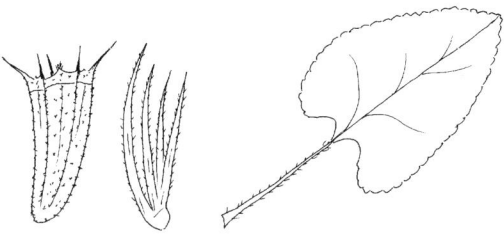

Phlomis canescens Regel

Acta. Horti. Petrop. 9:579 (1886)
Distribution in the wild:
   Afghanistan, former USSR, in forest-steppe, and sub-alpine zones, on
   stony slopes.
Flowering in the wild;   June–August
canescens means with short hairs that give the leaves a whitish appearance.

Herbaceous perennial 50–80 cm. Radical leaves ovate, deeply cordate at base, obtusish,
crenate at margin, 11–16 × 5–10 cm; petioles 15–25 cm. Cauline leaves similar 8 × 6 cm;
petioles 2–4 cm. Floral leaves oblong-ovate, acuminate, 7 ×3–4 cm, all except the uppermost
covered with small stellate hairs; petioles 0.3–0.5 cm to sessile. Few whorls. Bracteoles
filiform-subulate, 10–17 mm, connate at base in twos or free, covered with stellate, monoradial
and simple hairs. Calyx tubular-campanulate, 10–15 mm, teeth notched or rounded with a
point 2–4 mm long, covered with stellate and monoradial hairs. Corolla lilac-rose. Nutlets
hairless or with sparse, short hairs at apex.

Phlomis koraiensis Nakai ✧

Bot. Mag. (Tokyo) 31: 106 (1917)
Distribution in the wild:
   N. Korea, in alpine meadows, 2,200 m.
koraiensis means of Korea.

Herbaceous perennial to 45 cm. Basal leaves broadly ovate, upper surface rugose, lower
surface densely stellate hairy, deeply cordate at base, crenate at margin, apex obtuse or
acute, c. 14 × 12 cm; petiole 8–11.5 cm. Stem leaves cordate, crenate at margin, 5.5–8 × c.
5cm; petiole c. 2.5 cm. Floral leaves ovate to lanceolate, shallowly cordate to broadly cuneate
at base, irregularly dentate-crenate or shallowly serrate, 2.5–4.5 × 0.7–2.7 cm; short petiolate

to subsessile. Whorls c. 8 flowered. Bracteoles bristle-like, densely stellate puberulent and ciliate, 9–11 mm. Calyx campanulate, densely stellate puberulent, 11–12 mm; teeth broad at base, bearded, 2–3 mm. Corolla red-purple, c. 22 mm. Nutlets hairless.

Phlomis macrophylla Wall. ex Benth. ✦

Pl. As. Rar. 1:62 (1830)
Synonym:
Phlomoides macrophylla (Wall. ex Benth.) Kamelin & Makhmedov
Distribution in the wild:
Punjab, Himalaya (Uttar Pradesh to Bhutan), India, on open slopes, 3,100–4,100 m.
Flowering in the wild: July–August
macrophylla means with large leaves.
Seems to take several years to settle down and flower.

Herbaceous perennial to 2m. Heart-shaped, toothed, hairy leaves 10–20 cm long; petioles nearly as long. Bracteoles awl shaped, rigid, ciliate with spiny tips. Calyx with spine tipped lobes about one third as long as the calyx tube, often purple with few scattered hairs.

Phlomis maximowiczii Regel

Acta Hort. Petrop. 9:594 t.10, fig. 18 (1886)
Distribution in the wild:
China, former USSR, in deciduous and mixed woods, forest margins and river banks.
Flowering in the wild: July–August
maximowiczii is in honour of Carl Ivanovich Maximowicz (1827-1891)

Herbaceous perennial 80–100 cm. Basal leaves ovate, upper surface sparsely minute hispid, lower surface sparsely stellate-pannose, shallowly cordate at base, serrate or dentate at margin, acuminate at apex, 9–15 × 8–10 cm; petiole 7–9 cm. Upper stem leaves much reduced. Lower floral leaves ovate-lanceolate, subsessile. Upper floral leaves 2–3 × 1–2 cm, entire or dentate at margin. Whorls many flowered, separate, on stems branched towards apex, peduncle 1–2 mm. Bracteoles lanceolate, 9–10 mm, margin ciliate. Calyx tubular, slightly enlarged at apex, 8–10 mm, spreading bristly on veins, teeth truncate, puberulant except for apical tuft inside, apex spinescent. Corolla reddish, c. 2 cm. Nutlets pubescent.

Phlomis melanantha Diels

Notes Roy. Bot. Gard. Edinb. 5:242 ((1912)
Synonyms:
Phlomoides melanantha (Diels) Kamelin & Makhmedov
Distribution in the wild:
China (Sichuan, Yunnan), in *Picea* forests, mixed forests, grasslands, at 3,000–3,300m.
Flowering in the wild: June–September
melanantha means dark flowered.

Herbaceous perennial 60–90 cm with thick woody roots. Stem leaves broadly ovate to ovate-oblong, cordate at base, acute to long acuminate at apex, serrate-dentate to dentate at margin, 4.5–12 × (1.9-)2.5–9.5 cm; petioles 1.2–6 cm. Floral leaves similar to cauline leaves. Numerous whorls, many flowered. Bracteoles rigid, subulate, spinescent, partly ciliate, 1–1.2 cm. Calyx purple, c. 12 × 6 mm; teeth double toothed at margin, spines 2–3 mm. Corolla purple red to black crimson. Nutlets hairless.

P. melanantha Diels **var. angustifolia** C.Y. Wu
P. melanantha Diels **var. angustifolia f. pallidior** C.Y.Wu
*Flora Reipubl. Popul. Sin.*65(2):597 (1977)

Phlomis oreophila Kar. & Kir. ✦✧

Bull. Soc. Nat. Mosc. 15:426 (1842)
Synonym:
  Phlomoides oreophila (Kar. & Kir.) Adylov, Kamelin, Makhmedov
Distribution in the wild:
  China (Xingiang), Former USSR (Central Asia, Mongolia, Siberia),
  from the forest steppe and forest zones up to the subalpine zone, on
  stony slopes, grassy slopes, at 2100–3000 m.
Flowering in the wild: June–August
oreophila means mountain loving.

Herbaceous perennial 30–80 cm. Stringlike roots. Basal leaves ovate to broadly ovate, acuminate, base cordate, crenate at margin, 6.5–13 × 5–10 cm; petiole 6–14 cm. Stem leaves similar or circular, 6–11 × 3.2–7 cm. Floral leaves ovate-lanceolate or lanceolate-linear, 3–6 × 0.5–2 cm; upper side of leaves with scattered simple hairs, lower side densely stellate and fascicled hairs. Up to 5 whorls per stem of many flowers. Bracteoles thin, 10–13 mm, filiform-subulate with long spreading hairs sometimes glandular villous. Calyx tubular, 13–15 cm, covered with stellate and long simple hairs, densely on the veins, teeth broad-ovate or rounded, deeply notched, apex subulate-acuminate, with a point 2–2.5 mm. Corolla c. 24 mm, lilac to purple. Nutlets stellate hairy at apex.

P. oreophila Kar. & Kir. **var. oreophila** ✦✧
Bracts densely villous or sometimes intermixed with glandular villous
hairs; calyx stellate puberulant, finely villous on veins outside.
P. oreophila Kar. & Kir. **var. evillosa** C. Y. Wu
Bracts and calyx stellate pannose.

Phlomis pratensis Kar. & Kir. ✦

Bul. Soc. Nat. Mosc. 15:426 (1842)
Synonym:
  Phlomoides pratensis(Kar. & Kir.) Adylov, Kamelin, Makhmedov
Distribution in the wild:
  China, Former USSR (Central Asia). Mountains, in the woodland and
  scrub zone.
Flowering in the wild: July–August
pratensis means of the meadows.

Herbaceous perennial to 75 cm. Basal and lower stem leaves ovate to ovate-oblong, cordate at base, obtusely toothed, 10–17 × 3.5–15 cm; petioles 3–22 cm. Upper stem leaves similar, smaller; petiole 1–3 cm. Upper floral leaves ovate-oblong, with rounded or acute teeth; sessile. Upper side of leaves simple and stellate hairs, lower side with long multi-articulate hairs on ribs and scattered stellate between. Floral leaves ovate-oblong, sessile, dentate to subentire at margin. Several whorls per stem (candelabra branched), many flowered; whorls on short or elongated peduncles. Bracteoles connate, thickish, linear-subulate, 8–15 mm covered with stellate and fascicled hairs. Calyx tubular 10–15 mm, thick veined with simple and monoradial hairs; teeth notched, with point 1.5–2 mm from notch. Corolla purple to pink, 15–30 mm. Nutlets hairless.

P. pratensis Kar. & Kir. **f. glabra** Knorr.
A form where the leaves are hairless beneath except for simple hairs on the veins.

Phlomis rotata Benth. ex Hook. f.

Fl. Br. India 4:694 (1885)
Synonym:
   Lamiophlomis rotata (Benth. ex Hook. f.) Kudo
Distribution in the wild:
   Eastern Himalayas (Nepal to Bhutan), Tibet, SW China, at 4,700 m.
rotata means wheel-shaped, almost flat and refers to the leaves.
This species is now considered a *Lamiophlomis* to distinguish its unusual features. It is monocarpic (dies after flowering). Leaves of young plants (1–2 years old) are narrowly oblong or spathulate with petioles longer than the rugose laminas. At the flowering stage the leaves become broadly oblong or ovate to nearly circular with a basal rosette of leaves pressed flat to the ground. Probably more suited to an Alpine specialist grower.

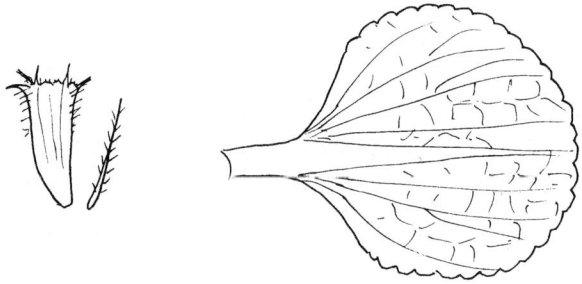

Leaves kidney-shaped 4–13 cm across, wrinkled and deeply veined above, toothed margin; very broad woolly petiole. Calyx c. 8mm, hairy, with tiny bristle-pointed teeth; bracteoles with spiny tips. Whorl borne on a stout stem to 15 cm. Flowers c. 1.5 cm, lower lip nearly entire. Corolla lilac.

P. rotata Benth. ex Hook. f. **subsp. bhutanica** R.A. Clement
Distribution in the wild: Endemic to Bhutan

Leaves ovate to broadly ovate, 3.6–8.2 × 3.1–6.8 cm, upper surface hairs simple (not stellate). One to two whorls. Calyx teeth shorter, 1.5 mm. Corolla lower lip slightly shorter than upper lip and hairless on upper surface.

Phlomis setigera Falc. ex Benth.

D.C. Prodr. 12:543 (1848)
Synonym:
   Phlomoides setigera (Falc. ex Benth.) Kamelin & Makhmedov
Distribution in the wild:
   Kashmir to Bhutan, Tibet, India at 2,400 to 3,600 m. in forest clearings.
Flowering in wild: June–July
setigera means bearing bristles and refers to the bracteoles

Herbaceous perennial to 2m. Heart shaped leaves with short petioles, 1–2.5 cm. Whorls many flowered. Bracteoles short, rigid, covered in bristles, tips spinous. Calyx 8–13 mm, hairy; teeth erect, awl shaped, half as long as calyx. Corolla pink to mauve. Nutlets 4mm long.

Phlomis tatsienensis Bureau & Franchet ✦✧

Morot, Journ. De Bot. 5:149 (1891)
Synonym:
    Phlomoides tatsienensis (Bureau & Franchet) Kamelin & Makhmedov
Distribution in the wild:
    China (Sichuan, Yunnan), on grassy slopes and forests at 2500-3400 m.
Flowering in the wild: May–July
tatsienensis means of Tatsienlu (K'angding) in western China (Tibet)

Herbaceous perennial 30–100 cm. Roots thick. Stem leaf lamina ovate, , stellate-hairy, cordate to rounded at base, acute to long acuminate at apex, crenate to serrate at margin, 5–23 × 2.5–17 cm; petiole 1–13 cm. Floral leaves 1–3 × 0.5–1.3 cm, serrate at margin; petiole 0.3–0.5 cm. Many whorls per stem, 6–14 flowered. Bracteoles few , linear, herbaceous, 3–6 mm, stellate pubescent. Calyx tubular-campanulate, c. 8 × 5 mm, stellate pubescent, teeth double toothed, tufted hairy spines c. 1.5 mm. Corolla white with purplish lower lip, c. 13 mm. Nutlets hairless.

P. tatsienensis Bureau & Franchet **var. tatsienensis** ✦✧
Bracteoles and calyx stellate pubescent only.
P. tatsienensis Bureau & Franchet **var. hirticalyx** Hand.–Mazz.
Fl. Yunnanica 1:612 (1977)
Distribution in the wild: China (Yunnan)
Synonyms: *P. souliei* Lév.
    P. franchetiana Hand.–Mazz.
Bracteoles and calyx densely stellate pubescent, spreading hirsute.

Phlomis tuberosa L. ✦✧

Sp. Pl. 586 (1753)
Illust.: *Bot. Mag.* t. 1555 (1813)
Colour Plates: V (i), (ii)
Synonyms:
    P. sythica Klokov & Sost.
    Phlomoides tuberosa (L.) Moench
Distribution in the wild:
    Hungary, former Yugoslavia, Bulgaria, Greece, Turkey, Iran, China and
    former USSR, on dry stony slopes, fallow fields, steppe and meadows.
Flowering in the wild: June–July
tuberosa means tuberous.
This plant has large underground tubers which can be cooked and eaten. It develops large coarse, green, arrow shaped leaves with long petioles. The flower stems are often bright purple and from three to five feet high; they can be simple or branched. The three to ten flower whorls on each stem each have may small purplish flowers. Hardy to –20°C at least.

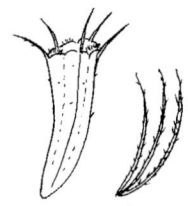

Herbaceous perennial 40–150 cm. Roots string like with large tubers. Basal leaf laminas green, papery, triangular, auriculate-cordate or sagittate at base, crenate or dentate at margin, 5–30 × 5–15 cm; petiole 4–30 cm. Floral leaves lanceolate, 5 × 2–2.5 cm, sharply serrate-dentate and sessile. Floral stems often purple and almost hairless, candelabra branched, or simple. 3–10 whorls per segment, many flowered. Whorls 4–5 cm across. Bracteoles linear subulate, branched in threes 8–13 mm, hairless or hairy, ciliate. Calyx tubular-campanulate, 8–13 mm, hairless except for bristles near teeth; teeth to 3.5 mm. Corolla 12–20 mm, pink to purple.

P. tuberosa 'Amazone' is an allegedly taller growing selection, but many of the forms of the species in cultivation are its equal in height in the same soil.

Phlomis umbrosa Turcz.

Bull. Soc. Imp. Nat. Mosc. 13:76 (1840)

Colour Plate: IV (v)

Distribution in the wild:

China, in forests, grassy slopes, streamsides, thickets, wet areas, at 700–3200 m.

Flowering in the wild: June–September

umbrosa means shade loving. However, it grows well in full sun in the UK.

Herbaceous perennial 50–150 cm. Leaf circular-ovate to ovate-oblong, pilose and/or stellate, cordate to rounded at base, serrate-dentate to irregularly crenate at margin, acute to acuminate at apex, 5.2–12 × 2.5–12 cm; petiole 1–12 cm. Floral leaves coarsely serrate-dentate at margin, 1–3.5 × 0.6–2 cm; petiole 2–3 mm. Bracteoles purple-red. Numerous whorls, 4–8 flowered, on much branched purple-red stems. Calyx tubular, 8–10 × 3–3.5 cm, densely stellate hairy, teeth small. Corolla reddish to purple-red, rarely white with red spots on lower lip. Nutlets hairless.

P. umbrosa Turcz. var. *umbrosa*

Leaf lamina papery, orbicular-ovate to ovate-oblong, serrate-dentate or irregularly crenate at margin, terminal tooth not very long. Bracts rigid, linear-subulate, mostly longer than calyx. Whorls inconspicuously pedunculate. Calyx c. 10 × 3.5 cm, stellate puberulent except sometimes pilose on veins. Hardy to –15°C.

P. umbrosa Turcz. var. *australis* Hemsley

J. Linn. Soc. Bot. 26:306 (1890)

Colour Plate: IV (v)

This variety is easily distinguished from the species by the truncated leaf apex. The upper part of the calyx is often purple. It is a good garden plant with nice flowers and habit.

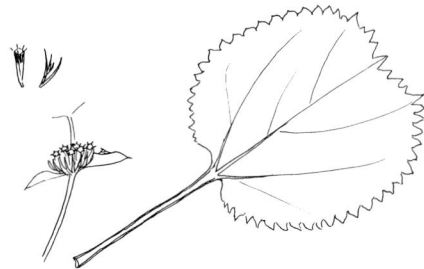

Leaves long petiolate, lamina membranous, crenate-serrate at margin, terminal tooth sometimes very long. Bracteoles soft, linear lanceolate, slightly shorter than the calyx. Hardy to –15°C

P. umbrosa Turcz. **var. latibracteata** Sun ex C.H. Hu
Acta Phytotax. Sin. 11:46 (1966)

Bracteoles linear-oblong to obovate-oblong, much shorter than the calyx, 5 – 7 × 1.8 – 2.5 mm, entire at margin.

P. umbrosa Turcz. **var. ovalifolia** C. Y. Wu
Fl. Reipubl. Popularis Sin. 65(2):601 (1977)
The epithet *ovalifolia* means oval leaved.

Lamina ovate, subcuneate to shallowly cordate at base, stellate pubescent, rarely abaxially stellate-tomentose. Calyx densely stellate pubescent, teeth longer.

P. umbrosa Turcz. **var. stenocalyx** (Diels) C.Y. Wu
Fl. Reipubl. Popularis Sin. 65 (20:477 (1977)
Synonym:
    P. stenocalyx Diels
stenocalyx means with a narrow calyx.

Leaves papery. Whorls conspicuously pedunculate, loose. Bracteoles very slender. Flowers pedicellate. Calyx c. 8 × 3 mm.

Phlomis urodonta M. Popov ✦

Nouv. Mém. Soc. Nat. Mosc. 19:40 (1941)
Synonym:
    Phlomoides urodonta (Popov) Adylov, Kamelin, Makhmedov
Distribution in the wild:
    Former USSR (Central Asia), in the upper part of the woodland and scrub zone, in scrub.
Flowering in the wild: June–July
urodonta means tail-like teeth, referring to the bracteoles

Herbaceous perennial 60–70 cm. Basal leaves ovate, cordate at base, crenate at margin, 15–16 × 6–7 cm; petiole 16–17 cm. Stem leaves similar, smaller; petiole to 3 cm. Floral leaves oblong, entire at margin, upper surface hairless, lower surface sparsely hairy on veins; sessile. Many whorls, many flowered on whitish stems, purplish towards top, palmately branched and hairless except for long hairs under whorls. Bracteoles numerous, lanceolate, as long as or slightly shorter than calyx. Calyx coriaceous, tubular-campanulate, prominently veined, the veins and margin of teeth with long articulate and simple hairs; calyx teeth broad-ovate, 3–4 mm, with a point 5–8 mm. Corolla straw-yellow or creamy-pink. Nutlets short haired at apex.

Appendix 1—*Phlomis* Distribution by Country

The short descriptions included with each plant are taken from Floras such as those for Turkey, USSR, China, Iran, Cyprus etc. It should be noted however that the details vary for some plants in different countries. For example, *Phlomis fruticosa* is described in the Flora of the USSR as an orange flowered shrub only 25–45 cm in height. Species discussed in the main text, are shown in **bold** text. The page number in the margin corresponds to where a description will be found in this book. Author names are given in full in this index.

Aegean Islands
20   ***P. cretica*** C. Presl   Yellow flowered shrub to 45 cm
24   ***P. grandiflora*** H.S.Thompson Yellow flowered shrub to 200 cm
32   ***P. lycia*** D.Don Yellow flowered shrub to 150 cm

Afghanistan
62   ***P. bracteosa*** Royle ex Bentham   Rose/violet herb, 20–100 cm
42   ***P. bucharica*** Regel   Yellow to brown red herb 40–60 cm
42   ***P. cancellata*** Bunge   White flowered herb, 20–35 cm
63   ***P. canescens*** Regel   Lilac flowered herb, 50–75 cm
43   ***P. cashmeriana*** Royle ex Bentham   Lilac/purple herb to 80 cm
       P. spectabilis Falconer ex Bentham   Rosy-purple herb, 30–200 cm
       P. stewartii J.D. Hooker   Dusky-pink herb, 30–45 cm
       P. trineura K.H.Rechinger   Magenta herb to 60 cm

Albania
23   ***P. fruticosa*** Linnaeus   Yellow flowered shrub to 200 cm
46   ***P. pungens*** Willdenow   Purple flowered herb, 30–75 cm

Algeria
39   ***P. bovei*** de Noé   Purple flowered herb to 100 cm
39   ***P. bovei*** subsp. ***maroccana*** Maire Purple flowered herb to 100cm
44   ***P. crinita*** Cavanilles   Brown/yellow flowered herb to 75 cm
       P. crinita subsp. *mauritanica* (Munby) Murbeck
46   ***P. pungens*** Willdenow   Purple flowered herb, 30–75 cm
35   ***P. purpurea*** subsp. ***caballeroi*** (Pau) Rivas Martínez

Balearic Islands
26   ***P. italica*** Linnaeus   Pale rose flowered shrub to 200 cm

Bulgaria
46   ***P. pungens*** Willdenow   Purple flowered herb, 30–75 cm
67   ***P. tuberosa*** Linnaeus   Purple flowered herb to 200 cm

Burma
62   ***P. breviflora*** Bentham   Whitish flowered herb to 70 cm
       P. burmanica Mukerjee

China

60   **P. agraria** Bunge   Reddish or white flowered herb, 40–60 cm
      P. albiflora Hemsley
60   **P. alpina** Pallas   Reddish flowered herb, 20–50 cm
      P. ambigua Handel-Mazzetti   Yellowish (reddish upper lip) fl. herb
61   **P. atropurpurea** Dunn   Dull purple flowered herb, 20–60 cm
61   **P. atropurpurea f. pallidor** C.Y. Wu   Pale coloured form.
61   **P. atropurpurea f. pilosa** C.Y. Wu   Long soft haired form.
61   **P. betonicoides** Diels   Reddish to white flowered herb, 30–80 cm
62   **P. betonicoides f. alba** C.Y. Wu   White flowered form.
      P. burmanica Mukerjee
      P. chingohensis C.Y. Wu   White flowered herb, 20–50 cm
      P. congesta C.Y. Wu   White flowered herb, 30–80 cm
      P. cuneata C.Y. Wu (Tibet)
      P. dentosa Franchet   Reddish flowered herb to 80 cm
      P. dentosa var. *glabrescens* Danguy
      P. esquirolii Léveillé
      P. fimbriata C.Y. Wu   Dark purple flowered herb, 20–30 cm
      P. forrestii Diels   Purplish flowered herb, 30–90 cm
      P. forrestii var. *taronensis* C.Y. Wu
      P. franchetiana Diels   Purple to white flowered herb, 60–90 cm
      P. franchetiana var. *leptophylla* C.Y. Wu
      P. franchetiana var. *aristata* C.Y. Wu
23   **P. fruticosa** Linnaeus   Orange flowered shrub, 25–45 cm.
      P. gracilis Hemsley
      P. hirticalyx (Handel-Mazzetti) C.Y. Wu
      P. inequalisepala C.Y. Wu Herb, 70–80 cm
      P. jeholensis Nakai & Kitagawa   White flowered herb about 75 cm
      P. kansuensis C.Y. Wu   Reddish flowered herb about 35 cm
      P. kawaguchi Murati (Tibet)
      P. longicalyx C.Y. Wu Herb.
      P. likiangensis C.Y. Wu   White or yellow flowered herb, 60–150 cm
      P. maximowiczii Regel (Manchuria) Reddish flowered herb, 80–100 cm
      P. medicinalis Diels   Purple-red or reddish herb, 20–75 cm
      P. megalantha Diels   Yellowish to white fl. herb, 15–45 cm
      P. megalantha var. *pauciflora* C.Y. Wu
64   **P. melanantha** Diels   Purple-red flowered herb, 60–90 cm
      P. milingensis C.Y. Wu & H.W. Li (Tibet)   Purple-red herb 15–40 cm
      P. mongolica Turczaninow (Mongolia)   Purple flowered herb, 40–70 cm
      P. mongolica var. *macrocephala* C.Y. Wu   Purple fl. herb to 15 cm
      P. muliensis C.Y. Wu   White flowered herb, 60–100 cm
65   **P. oreophila** Karelin & Kirilov   Purple flowered herb, 30–80 cm
65   **P. oreophila var. evillosa** C.Y. Wu
      P. ornata C.Y. Wu   Dark purple flowered herb, 40–60 cm
      P. ornata var. *minor* C.Y. Wu
      P. paohsingensis C.Y. Wu   Purplish flowered herb, 90–120 cm

P. pararotata Sun ex C.H. Hu   Red flowered herb about 35 cm

P. pedunculata Y.Z. Sun   White or purplish flowered herb, 50–80 cm

65   **P. pratensis** Karelin & Kirilov   Purple/ pink flowered herb to 75cm

P. pygmaea C.Y. Wu (Tibet)   Purple flowered herb about 5 cm

66   **P. rotata** Bentham ex J.D. Hooker   Lilac flowered herb to 15 cm

P. ruptilis C.Y. Wu   Yellow flowered herb.

P. setifera Bureau & Franchet   Yellow-white flowered herb, 30–50 cm

66   **P. setigera** Falconer ex Bentham (Tibet) Mauve flowered herb to 200cm

P. similis Cherneva

P. souliei Léveillé

P. strigosa C.Y. Wu   White flowered herb, 55–100 cm

P. szechuanensis C.Y. Wu   White flowered herb

P. taronensis C.Y. Wu

67   **P. tatsienensis** Bureau & Franchet White with purplish lower lip, 30–100 cm

67   **P. tatsienensis var. hirticalyx** Handel-Mazzetti

P. tibetica Marquand & Airy Shaw (Tibet) Purple or pink herb,18–50 cm

P. tibetica var. *wardii* Marquand & Airy Shaw.

67   **P. tuberosa** Linnaeus   Purple-red flowered herb, 40–150 cm

P. tuvinica Shreter (Mongolia)

68   **P. umbrosa** Turczaninow   Reddish to purple-red herb, 50–150 cm

68   **P. umbrosa var. australis** Hemsley

69   **P. umbrosa var. latibracteata** Sun ex C.H. Hu

69   **P. umbrosa var. ovalifolia** C.Y. Wu

69   **P. umbrosa var. stenocalyx** (Diels) C.Y. Wu

P. uniceps C.Y. Wu   Purple flowered herb to about 10 cm

P. wangii Hu & Tsai

P. younghusbandii Mukerjee (Tibet)   Herb 15–20 cm

Crete & Karpathos

20   **P. cretica** C. Presl   Yellow flowered shrub to 45 cm

23   **P. floccosa** D.Don   Yellow flowered shrub to 100 cm

23   **P. fruticosa** Linnaeus   Yellow flowered shrub to 200 cm

27   **P. lanata** Willdenow   Yellow flowered shrub to 60 cm

36   **P. sieberi** Vierhapper

33   **P. pichleri** Vierhapper   Yellow flowered shrub

20   **P. × commixta** K.H.Rechinger (*P. cretica* × *P. lanata*)

23   **P. × vierhapperi** K.H.Rechinger (*P. floccosa* × *P. pichleri*)

Crimea

P. hybrida Zelenetsky

46   **P. pungens** Willdenow   Purple flowered herb, to 30–75 cm

48   **P. taurica** Bunge   Purple flowered herb to 50 cm

67   **P. tuberosa** Linnaeus   Purple flowered herb to 150 cm

Cyprus

18   **P. brevibracteata** Turrill   Yellow flowered shrub, to 150 cm

21	*P. cypria* Post Yellow flowered shrub, 50–150 cm
21	*P. cypria* var. *occidentalis* Meikle
23	*P. fruticosa* Linnaeus Yellow flowered shrub, 100–150 cm
30	*P. longifolia* Boissier & Blanche Yellow fl. shrub to 100 cm
30	*P. longifolia* var. *bailanica* (Vierhapper) Huber–Morath
31	*P. lunariifolia* Smith Yellow flowered shrub, 100–200 cm

Egypt

17	*P. aurea* Decaisne Yellow flowered shrub to 90 cm
23	*P. floccosa* D.Don Yellow flowered shrub to 100 cm

France

45	*P. herba-venti* Linnaeus Purple flowered herb to 30–75 cm
50	*P. lychnitis* Linnaeus Yellow flowered herb to 65 cm

Greece

20	*P. cretica* C Presl Yellow flowered shrub to 45 cm
23	*P. fruticosa* Linnaeus Yellow flowered shrub to 200 cm
45	*P. herba-venti* Linnaeus Purple flowered herb to 30–75 cm
46	*P. pungens* Willdenow Purple flowered herb, 30–75 cm
57	*P. samia* Linnaeus Purple flowered herb to 150 cm
20	*P.* × *cytherea* K.H.Rechinger (*P. cretica* × *P. fruticosa*)

India (Himalayas)

62	*P. bracteosa* Royle Purple flowered herb 20-80 cm
62	*P. breviflora* Bentham Whitish flowered herb to 70 cm
43	*P. cashmeriana* Royle ex Bentham Lilac flowered herb to 90 cm
66	*P. rotata* Bentham ex J.D. Hooker
66	*P. setigera* Falconer ex Bentham Pink to mauve flowered herb to 200cm
	P. simplex Royle

Iran

37	*P. anisodonta* Boissier Pink flowered herb, 30–40 cm
38	*P. aucheri* Boissier Yellow flowered herb, 30–60 cm
62	*P. bracteosa* Royle ex Bentham Purple flowered herb, 20–50 cm
41	*P. bruguieri* Desfontaines Yellow flowered herb, 20–30 cm
42	*P. bucharica* Regel Yellow (upper lip dirty red) herb
42	*P. cancellata* Bunge White flowered herb, 20–35 cm
63	*P. canescens* Regel Rose-Lilac flowered herb, 50–80 cm
43	*P. cashmeriana* Royle ex Bentham Pink flowered herb to 40 cm
	P. caucasica K.H.Rechinger Yellow flowered herb, 30–50 cm
	P. chorassanica Bunge White flowered herb to 20 cm
22	*P. elliptica* Bentham Pink flowered shrub, 40–50 cm
	P. ghilanensis C. Koch Yellow flowered herb.
45	*P. herba-venti* Linnaeus Pink to purple-violet herb, 30–70 cm
46	*P. herba-venti* subsp. *lenkoranica* (Knorring) K.H.Rechinger
46	*P. herba-venti* subsp. *kopetdaghensis* (Knorring) K.H.Rechinger

P. kuegleriana Muschler
48   **P. kurdica** Rech.   Yellow flowered herb, 50–60 (–70) cm
P. lanceolata Boissier & Hohenacker
P. olivieri Bentham   Yellow flowered herb, 25–35 (–60) cm
P. pachyphylla K.H.Rechinger   Yellow flowered herb, 10–20 cm
54   **P. persica** Boissier   Pink flowered herb, 30–40 cm
P. polioxantha K.H.Rechinger   Yellow flowered herb, 20–30 cm
46   **P. pungens** Willdenow   Lilac flowered herb, 30–60 cm
56   **P. rigida** Labillardière   Pink flowered herb , 40–60 cm.
P. spectabilis Falconer ex Bentham   Pink flowered herb, 120–180 cm
P. stewartii J.D. Hooker   Pink flowered shrub, 30–45 cm.
P. trineura K.H.Rechinger   Dirty pink flowered herb to 45 cm
67   **P. tuberosa** Linnaeus   Pink flowered herb, 40–150 cm
42   **P. x pabotii** K.H.Rechinger *(P. bruguieri* x *P. pachyphylla)*
42   **P. x praetervisa** K.H.Rechinger *(P. bruguieri* x *P. kurdica)*
39   **P. x stapfiana** K.H.Rechinger *(P. aucheri* x *P. olivieri)*

Iraq

37   **P. anisodonta** Boissier   Pink flowered herb, 30-40 cm
38   **P. armeniaca** Willdenow   Yellow flowered herb to 60 cm
38   **P. aucheri** Boissier   Yellow flowered herb, 30–60 cm
P. brevilabris C.G. Ehrenberg
41   **P. bruguieri** Desfontaines   Yellow flowered herb, 20–80 cm
42   **P. cancellata** Bunge   White flowered herb, 20–35 cm.
P. elongata Handel-Mazzetti
45   **P. herba-venti** Linnaeus   Purple flowered herb 30–75 cm
48   **P. kurdica** K.H.Rechinger   Yellow flowered herb 60 cm plus
P. lanceolata Boissier & Hohenacker
P. olivieri Bentham   Yellow flowered herb, 25–35 (–60) cm
P. polioxantha K.H.Rechinger   Yellow flowered herb, 20–30 cm
46   **P. pungens** Willdenow   Lilac flowered herb, 30–60 cm
55   **P. rigida** Labillardière   Pink flowered herb , 40–60 cm
P. x praetervisa K.H.Rechinger *(P. bruguieri* x *P. kurdica)*

Israel & Jordan

40   **P. brachycodon** (Boissier) Zohary   Yellow flowered herb, 30–50 cm
48   **P. kurdica** K.H.Rechinger   Pale yellow flowered herb, 30–60 cm
54   **P. platystegia** Post   Yellow flowered herb, 50–100 cm
46   **P. pungens** Willdenow   Lilac flowered herb, 30–60 cm
58   **P. syriaca** Boissier   Yellow flowered herb, 50–70 cm
36   **P. viscosa** Poiret   Yellow flowered shrub, 80–150 cm

Italy

22   **P. ferruginea** Tenore   Yellow flowered shrub to
23   **P. fruticosa** Linnaeus   Yellow flowered shrub to 200 cm
45   **P. herba-venti** Linnaeus   Pink to Purple flowered herb 30-70 cm
27   **P. lanata** Willdenow   Yellow flowered shrub to 60 cm

Kashmir

43   ***P. cashmeriana*** Royle ex Bentham   Lilac/purple herb to 80 cm
      P. cashmirica Wells
66   ***P. setigera*** Falconer ex Bentham Pink to mauve flowered herb to 200cm

Korea (North)

63   ***P. koraiensis*** Nakai   Red-purple flowered herb about 45 cm

Lebanon & Syria

38   ***P. armeniaca*** Willdenow   Yellow flowered herb to 60 cm
17   ***P. aurea*** Decaisne   Yellow flowered shrub to 90 cm
      P. bertrami Post
40   ***P. brachycodon*** (Boissier) Zohary   Yellow flowered herb, 30–50 cm
41   ***P. brachycodon* subsp. *damascena*** Bornmüller
      P. brevilabris Boissier
41   ***P. bruguieri*** Desfontaines   Yellow flowered herb, 20–30 cm
43   ***P. capitata*** Boissier   Yellow flowered herb, 10–30 cm
      P. cordata Boissier et Kotschy
19   ***P. chrysophylla*** Boissier   Yellow flowered shrub to 100 cm
23   ***P. floccosa*** D.Don   Yellow flowered shrub to 100 cm
23   ***P. fruticosa*** Linnaeus   Yellow flowered shrub to 130 cm
45   ***P. herba-venti*** Linnaeus   Pink to purple-violet herb, 30–70 cm
      P. kotschyana Huber–Morath
48   ***P. kurdica*** K.H.Rechinger   Pale yellow flowered herb, 30–60 cm
27   ***P. lanata*** Linnaeus   Yellow flowered shrub to 60 cm
49   ***P. linearis*** Boissier et Balansa   Yellow herb, 10–40 cm
30   ***P. longifolia*** Boissier et Blanche   Yellow flowered shrub to 130 cm
52   ***P. nissolii*** Linnaeus   Yellow herb to 100 cm
      P. orientalis Miller
54   ***P. platystegia*** Post   Yellow flowered herb, 50–100 cm
46   ***P. pungens*** Willdenow   Purple flowered herb, 30–75 cm.
55   ***P. rigida*** Labillardière   Pink flowered herb, 40–60 cm
      P. shepardi Post
58   ***P. syriaca*** Boissier   Yellow flowered herb, 50–70 cm
36   ***P. viscosa*** Poiret   Yellow flowered shrub, 80–150 cm

Libya

23   ***P. floccosa*** D.Don   Yellow flowered shrub to 100 cm

Morocco

39   ***P. bovei*** de Noé   Purple flowered herb to 100 cm
39   ***P. bovei* subsp. *maroccana*** Maire   Purple flowered herb to 150cm
      P. purpurea subsp. *antiatlantica* (Peltier) Rivas-Martínez
35   ***P. purpurea* subsp. *caballeroi*** Pau
44   ***P. crinita*** Cavanilles   Brown/Yellow flowered herb to 75 cm
44   ***P. crinita* subsp. *mauritanica*** (Munby) Murbeck
46   ***P. pungens*** Willdenow   Lilac flowered herb, 30–60 cm
51   ***P. × composita*** Pau (*P. crinita* × *P. lychnitis*)

Nepal

62	**P. bracteosa** Royle ex Bentham	Purple flowered herb, 20–80 cm
62	**P. breviflora** Bentham	Whitish flowered herb to 75 cm
64	**P. macrophylla** Wall	Herb to 200 cm
66	**P. rotata** Bentham	Herb to 15 cm
66	**P. setigera** Falconer	Pink/mauve herb to 200 cm
	P. spectabilis Falconer	Herb, 30–200 cm
	P. tibetica Marquand & Shaw	Purple or pink herb, 18–50 cm

Pakistan

62	**P. bracteosa** Royle ex Bentham	Purple herb, 20–50 (–100)cm
43	**P. cashmeriana** Royle ex Bentham	Rose herb to 40–80cm
	P. spectabilis Falconer ex Bentham	Rose/Pink herb, 100–200 cm
	P. stewartii J.D. Hooker	Pink flowered shrub, 30–45 cm

Portugal

45	**P. herba-venti** Linnaeus	Rose flowered herb to 75 cm
50	**P. lychnitis** Linnaeus	Yellow flowered herb to 65 cm
34	**P. purpurea** Linnaeus	Purple flowered shrub to 200cm

Sardinia

23	**P. fruticosa** Linnaeus	Yellow flowered shrub to 200 cm

Sicily

23	**P. fruticosa** Linnaeus	Yellow flowered shrub to 200 cm
45	**P. herba-venti** Linnaeus	Purple flowered herb to 75 cm

Sinai

17	**P. aurea** Decaisne	
40	**P. brachycodon** (Boissier) Zohary (Doubtfully native)	
23	**P. floccosa** D.Don	Yellow flowered shrub to 100 cm

Spain

44	**P. crinita** Cavanilles	Yellow/brown flowered herb to 75 cm
45	**P. crinita var. malacitana** Pau	
45	**P. herba-venti** Linnaeus	Rose flowered herb to 75 cm
50	**P. lychnitis** Linnaeus	Yellow flowered herb to 65 cm
46	**P. pungens** Willdenow	Purple flowered herb, 30–75 cm.
34	**P. purpurea** Linnaeus	Purple flowered shrub to 200 cm
35	**P. purpurea subsp. almeriensis** (Pau) Losa & Rivas Goday	
51	**P. x almijarensis** Pau *(P. lychnitis × P. crinita)*	
51	**P. x composita** Pau *(P. crinita × P. lychnitis)*	
51	**P. x trullenquei** Pau *(P. crinita × P. lychnitis)*	
35	**P. xmargaritae** Silvestre & Aparicio *(P. x composita × P. purpurea)*	

Syria

see Lebanon & Syria

Tibet

see China

Tunisia

39	*P. bovei* de Noé	Purple flowered herb to 100 cm
44	*P. crinita* Cavanilles	Brown/Yellow flowered herb to 75 cm
44	*P. crinita* subsp. *mauritanica*	
23	*P. floccosa* D.Don	Yellow flowered shrub to 100 cm
46	*P. pungens* Willdenow	

Turkey

16	*P. amanica* Vierhapper	Yellow flowered shrub to 100 cm
37	*P. angustissima* Huber–Morath	Yellow herb to 60 cm
38	*P. armeniaca* Willdenow	Yellow flowered herb to 60 cm
17	*P. bourgaei* Boissier	Yellow flowered shrub to 150 cm
41	*P. bruguieri* Desfontaines	Yellow flowered herb to 80 cm
	P. brunneogaleata Huber–Morath	Yellow (brownish upper lip) to 65 cm
43	*P. capitata* Boissier	Yellow flowered herb, 10–30 cm
	P. carica K.H.Rechinger	Yellow flowered herb to 60 cm
19	*P. chimerae* Boissieu	Yellow flowered shrub to 30 cm
23	*P. fruticosa* Linnaeus	Yellow flowered shrub to 130 cm
24	*P. grandiflora* H. S. Thompson	Yellow flowered shrub to 200 cm
26	*P. grandiflora* var. *fimbrilligera* Hub.–Mor.	
	P. integrifolia Huber–Morath	Pink flowered herb, 20–35 cm
	P. kotschyana Huber–Morath	Yellow flowered herb, 15–30 cm
48	*P. kurdica* K.H.Rechinger	Yellow flowered herb to 60 cm
	P. lanceolata Boissier & Hohenacker	Yellow herb, 20–45 cm
28	*P. leucophracta* P.H.Davis & Huber–Morath	Yellow (brownish upper lip) shrub to 150 cm
49	*P. linearis* Boissier & Blanche	Yellow herb, 10–40 cm
30	*P. longifolia* Boissier & Blanche	Yellow shrub to 130 cm
30	*P. longifolia* var. *bailanica*	Yellow shrub to 130 cm
31	*P. lunariifolia* Smith	Yellow flowered shrub to 130 cm
32	*P. lycia* D.Don	Yellow flowered shrub to 150 cm
33	*P. monocephala* P.H.Davis	Yellow flowered shrub to 150 cm
52	*P. nissolii* Linnaeus	Yellow herb to 100 cm
53	*P. oppositiflora* Boissier & Haussknecht	Yellow flowered herb to 50 cm
	P. physocalyx Huber–Morath	Yellow flowered herb to 30 cm
46	*P. pungens* Willdenow	Purple flowered herb, 30–75 cm
47	*P. pungens* var. *hirta* Velenovsky	Purple flowered herb, 30–75 cm
47	*P. pungens* var. *hispida* Huber–Morath	Purple fl. herb, 30–75 cm
47	*P. pungens* var. *laxiflora* Velenovsky	Purple fl. herb, 30–75 cm
47	*P. pungens* var. *seticalycina* (Nab.) Huber–Morath	
56	*P. rigida* Labillardière	Pink flowered herb, 50–125 cm
56	*P. russeliana* (Sims) Bentham	Yellow flowered herb to 100 cm
57	*P. samia* Linnaeus	Purple flowered herb to 100 cm

P. sieheana K.H.Rechinger   Yellow flowered herb, 10–30 cm
P. sintenisii K.H.Rechinger   Yellow flowered herb, 20–35 cm
58   ***P. syriaca*** Boissier   Yellow flowered herb to 100 cm.
67   ***P. tuberosa*** Linnaeus   Purple or pink flowered herb to 150 cm
36   ***P. viscosa*** Poiret   Yellow flowered shrub to 130 cm
29   ***P.* × *alanyense*** Huber–Morath (*P. leucophracta* × *P. lunariifolia*)
38   ***P.* × *bornmuelleri*** K.H.Rechinger (*P. armeniaca* × *P. nissolii*)
32   ***P.* × *cilicica*** Huber–Morath (*P. lunariifolia* × *P. monocephala*)
50   ***P.* × *kalanensis*** Huber–Morath (*P. linearis* × *P. oppositiflora*)
49   ***P.* × *melitenense*** Huber–Morath (*P. kurdica* × *P. oppositiflora*)
18   ***P.* × *mobullensis*** Huber–Morath (*P. bourgaei* × *P. grandiflora*)
25   ***P.* × *muglensis*** Huber–Morath (*P. grandiflora* × *P. lycia*)
38   ***P.* × *rechingeri*** Huber–Morath (*P. armeniaca* × *P. carica*)
18   ***P.* × *termessi*** P.H.Davis (*P. bourgaei* × *P. lycia*)
50   ***P.* × *tunceliensis*** Huber–Morath (*P. kurdica* × *P. linearis*)

Former USSR

60   ***P. agraria*** Bunge   Pink or white flowered herb, 40–60 cm
 P. alaica Knorring   Creamy-pink flowered herb, 70–85 cm
60   ***P. alpina*** Pallas   Pink flowered herb, 20–50 cm
 P. angrenica Knorring   Creamy-pink flowered herb, 50–60 cm
 P. betonicifolia Regel   Lilac flowered herb, 40–60 cm
 P. brachystegia Bunge   Lilac flowered herb to 100 cm
42   ***P. bucharica*** Regel   Yellow flowered herb, 40–60 cm
42   ***P. cancellata*** Bunge   Yellow or white flowered herb, 20–30 cm
63   ***P. canescens*** Regel   Lilac flowered herb, 50–75 cm
43   ***P. cashmeriana*** Royle ex Bentham   Lilac/purple herb to 80 cm
 P. caucasica K.H.Rechinger   Yellow flowered herb, 20–65 cm
 P. cordifolia (Regel) Adylov, Kamelin et Makhmedov
 P. cyclodon Knorring   Pink flowered herb, 45–75 cm
 P. drobovii M. Popov   Lilac-rose flowered herb, 20–25 cm
 P. dszumrutensis Afanassiev   Lilac flowered herb, 50–55 cm
 P. ferganensis M. Popov   Whitish flowered herb to 80 cm
45   ***P. fruticetorum*** Gontscharov   Pink flowered herb, 50–80 cm
23   ***P. fruticosa*** Linnaeus   Orange flowered shrub 25–45 cm
46   ***P. herba-venti* subsp. *kopetdagensis*** (Knorring) K.H.Rechinger
46   ***P. herba-venti* subsp. *lenkoranica*** (Knorring) K.H.Rechinger
 P. hybrida Zelenetzky   Pink flowered herb, 35–60 cm
 P. hypanica Schostenko   Herb 20–60 cm
 P. hypoleuca Vvedensky   Pinkish-lilac flowered herb, 40–60 cm
 P. knorringiana M. Popov   Herb to 100 cm
 P. linearifolia Zakirov   Lilac flowered herb, 25–35 cm
 P. maeotica Schostenko   Herb 50–85 cm
 P. majkopensis (Novopokrovsky) Grossheim   Pink herb, 40–50 cm
 P. maximowiczii Regel   Pink flowered herb, 80–100 cm
 P. nubilans Zakirov   Creamy-pink flowered herb
53   ***P. olgae*** Regel   Purple flowered herb, 30–50 cm

65	**P. oreophila** Karelin et Kirilov Lilac flowered herb, 30–80 cm
	P. ostrovskiana Regel Pale lilac flowered herb, 40–90 cm
65	**P. pratensis** Karelin et Kirilov Pink flowered herb to 75 cm
	P. pseudopungens Knorring Pinkish-violet flowered herb, 30–75 cm
	P. puberula Krylov & Sergievskaja Lilac-pink herb, 25–35 cm
46	**P. pungens** Willdenow Purple flowered herb, 30–75 cm
55	**P. regelii** M Popov Lilac-pink flowered herb, 30-50 cm
	P. sythica Klokov & Schost. Lilac-rose flowered herb, 30–80 cm
	P. sewerzowii Regel Pink flowered herb, 30–45 cm
	P. spinidens Nevski Purple-rose flowered herb, 50–65 cm
	P. stepposa Klokov
48	**P. taurica** Hartwiss ex Bunge Purple flowered herb, 40–50 cm
	P. tenuis Knorring Creamy flowered herb, 20–25 cm
	P. terkestana Bunge
59	**P. thapsoides** Bunge Lilac flowered herb, 30–60 cm
	P. tomentosa Regel Yellow flowered herb to 40 cm
	P. tschimganica Vvedensky Lilac flowered herb, 45–50 cm
67	**P. tuberosa** Linnaeus Pink or lilac flowered herb, 40–150 cm
	P. tytthaster Vvedensky Pink flowered herb, 40–50 cm
69	**P. urodonta** M. Popov Straw-yellow or creamy-pink fl. herb, 60–70 cm
	P. vavilovii M. Popov Purple flowered herb, 40–60 cm
	P. zenaidae Knorring Creamy flowered herb, 25–45 cm

Former Yugoslavia

23	**P. fruticosa** Linnaeus Yellow flowered shrub to
46	**P. pungens** Willdenow Lilac flowered herb, 30–60 cm
57	**P. samia** Linnaeus Purple flowered herb to 150 cm
67	**P. tuberosa** Linnaeus Purple flowered herb to 150 cm

Appendix 2—*Phlomis* species re-classified over the years

P. africana Beauv. —Now *Leonotis pallida*
P. alba Blanco. —Now *Anisomeles ovata*
P. alba Forsk. —Now *Ballota forskahlei*
P. aspera Willd. —Now *Leucas aspera*
P. biflora Roxb. —Now *Leucas procumbens*
P. bracteata Herb. Madr. ex Wall. —Now *Leucas capitata*
P. calycina Roxb. —Now *Roylea elegans*
P. caribaea Jacq. —Now *Leucas martinicensis*
P. cephalotes Koen. ex Roxb. —Now *Leucas capitata*
P. chinensis Blume —Now *Leucas javanica*
P. chinensis Retz. —Now *Leucas chinensis*
P. chinensis Wight ex Wall. —Now *Leucas marrubioides*
P. ciliata Heyne. ex Wall. —Now *Leucas ciliata*
P. clandescina Bory & Chaub. —Now *Sideritis theezans*
P. decemdentata Willd. —Now *Leucas flaccida*
P. diffusa Rottl. ex Hook f. —Now *Leucas diffusa*
P. eriostoma Heyne ex Hook f. —Now *Leucas eriostima*
P. esculenta Roxb. —Now *Leucas aspera*
P. flaccida Steud. —Now *Leucas flaccida*
P. flavescens Boerl. —Now *Anthocoma flavescens*
P. glabrata Vahl —Now *Leucas glabrata*
P. gracilis Salisb. —Now *Leucas zeylanica*
P. hirsuta Wight ex Wall. —Now *Leucas dimidiata*
P. hirta Herb. Madr. ex Wall. —Now *Leucas helianthemifolia*
P. hirta Heyne ex Roth —Now *Leucas hirta*
P. iberica Vis. ex Boiss. —Now *Eremostachys laciniata*
P. inderiensis Schmalh. —Now *Eremostachys tuberosa*
P. indica Herb. Madr. ex Wall. —Now *Leucas dimidiata*
P. indica Linnaeus —Now *Leucas indica* R. Br.
P. javanica Prain. —Now *Leonurus javanicus* Blume
P. laciniata Linnaeus —Now *Eremostachys laciniata*
P. laciniata Willd. ex Ledeb. —Now *Eremostachys gymnocalyx*
P. lanigera Siev. —Now *Eremostachys gymnocalyx*
P. leonitis Linnaeus —Now *Leonotiis ocymifolia*
P. leonurus Linnaeus —Now *Leonotis leonorus*
P. linifolia Roth. —Now *Leucas lavandulaefolia*
P. martinicensis Sw. —Now *Leucas martinicensis*
P. *membranifolia* Ridley —Now *Gomphostemma membranifolia*
P. micrantha Burch. —Now *Stachys burchellii*
P. mollis Schum. & Thonn. —Now *Leucas martinicensis*
P. molllluccana Roxb. —Now *Leucas flaccida*
P. moluccoides Vahl. —Now *Otostegia scariosa*
P. montana Koen. ex Roxb. —Now *Leucas chinensis*
P. montana Roth. —Now *Leucas montana*

P. montana Rottl. ex Wall. —*Now leucas marrubioides*
P. nepetaefolia Linnaeus —Now *Leonotis nepetaefolia*
P. nutans Roth. —Now *Leucas nutans*
P. obliqua Buch.-Ham. ex Hook f. —Now *Leucas aspera*
P. oblongifolia Kuntze —Now *Leonurus oblongifolius* Blume
P. oblongifolia Prain. —Now *Leonurus oblongifolius* Blume
P. octodentata Stokes —Now *Leucas zeylanica*
P. pallida Schum. & Thonn. —Now *Leonotis pallida*
P. parvifolia Burch. —Now *Stachys integrifolia*
P. pilosa Herb. Madr. ex Wall. —Now *Leucas nepetaefolia*
P. pilosa Roxb. —Now *Leucas pilosa*
P. pluckenetii Roth. —Now *Leucas aspera*
P. repens Willd. ex Benth. —Now *Leucas decumbens*
P. rodontia Buch.-Ham. ex Wall. —Now *Leucas montana*
P. sibirica Medic. —Now *Leonurus sibirica*
P. speciosa Salisb. —Now *Leonotis leonurus*
P. stricta Heyne ex Hook f. —Now *Leucas stricta*
P. tomentosa Herb. Madr. ex Wall. —Now *Leucas marrubioides*
P. urticifolia Vahl. —Now *Leucas urticaefolia*
P. zeylanica Blanco. —Now *Leucas aspera*
P. zeylanica Linnaeus —Now *Leucas zeylanica*
P. zeylanica Roxb. —Now *Leucas lavandulaefolia*

Appendix 3—Glossary of botanical terms to describe *Phlomis*

Note the difference between for example *linear-lanceolate* (which means the leaves are of unvarying shape somewhere between linear and lanceolate) and *linear to lanceolate* (which means some leaves are linear and others lanceolate.)

Acuminate: Terminating very gradually in a point.

Adpressed: Lying flat against. Often used to describe the bracteoles lying tight against the calyx. Compare with **Divergent**

Attenuate: (of the apex or base) Tapering finely and concavely to a long drawn out point.

Bract: The floral leaves, i.e. the pair of leaves subtending the inflorescence.

Bracteole: The small leaf like organs (few to many) subtending the flowers that are themselves subtended by the bracts (floral leaves).

Elliptic: acute at each end. Length: breadth 2:1 to 3:2 with sides curved equally from middle.

Canescent: Densely covered with short grey-white pubescence.

Cauline leaves: leaves attached to the stem.

Ciliate: Having fine hairs resembling an eyelash, at the margin.

Congested: Used to describe a flower stem where the whorls are close together. Compare **Distant**.

Connate: Joined together e.g. *Phlomis* bracteoles are often joined together at their base.

Cordate: Having two equal more or less rounded lobes forming a deep sinus at base.

Crenate: Refers to leaf margins with convex teeth.

Crenulate: Minutely crenate

Cuneate: Wedge-shaped. With straight sides converging at the base.

Decurrent: Where the base of a leaf gradually extends down to join the petiole such that one cannot identify where the leaf blade ends and the leaf stalk or petiole begins. As in the leaves of *P. lychnitis*.

Dentate: Having sharp teeth with concave (or straight) edges (usually pointing directly outwards). Compare **Serrate** and **Serrulate**.

Denticulate: Minutely dentate.

Distant: Used to describe a flower stem where the whorls are widely spaced. Compare **Congested**.

Divergent: Used to describe the bracteoles where they are held loosely away from the calyx; the opposite of **Adpressed**

Floccose: Covered with dense (usually stellate) hairs, which fall away in little tufts when rubbed. Where this occurs on *Phlomis* leaves there is usually a dense layer of strongly held smaller stellate hairs beneath a layer of loosely held stalked stellate hairs. Where it occurs on stems, it usually leaves the stems bare of all hairs.

Floral leaves: the leaves subtending the inflorescence—also called bracts.

Galeate: helmet shaped—referring to the shape of the upper lip of *Phlomis*, section *Phlomis*, as opposed to section *Phlomoides*.

Glabrous: Hairless.

Glabrescent: Nearly glabrous or minutely and invisibly pubescent.

Glandular: Covered with hairs bearing glands upon their tips—see p. 86.

Hispid: Covered in coarse rigid erect hairs or bristles, harsh to the touch.

Lamina: Leaf blade.

Lanate (or Lanose): Felted or woolly. Usually stellate-lanate in *Phlomis*. Little different in practice (in descriptions of *Phlomis*) from **Pannose** and **Tomentose** due to being author dependent.

Lanceolate: spear shaped, tapering towards the apex. Length: breadth about 3:1. Compare **Oblanceolate**.

Membranous: Thin textured, soft and flexible.

Mucro: An abrupt, sharp, terminal spur, spine or tip.

Mucronate: (of a leaf apex) Terminating suddenly with an abrupt spur or spine arising wholly from the midrib.

Petiole: the leaf stem which attaches the leaf blade or lamina to the main stem.

Reticulate: Netted. i.e. with a close or open network of veins, ribs or colouring.

Oblanceolate: lanceolate reversed. i.e. broadest beyond the middle point. Compare **Lanceolate.**

Obovate: Ovate reversed. i.e. broadest beyond the middle point. Compare **Ovate.**

Oblong: Elliptical, obtuse at each end.

Obtuse: Blunt. Terminating gradually in a rounded end (i.e. rounded enough for an angle of 90° to be placed inside it.

Ovate: Oblong or elliptical, broadest at the lower end, so as to resemble the longitudinal section of an egg. Length: breadth 2:1 to 3:1, broadest below the middle. Compare **Obovate.**

Pannose: Felt-like in texture, being densely covered in woolly hairs. In practice (in descriptions of *Phlomis*) little different from **Lanate** (or **Lannose**) or **Tomentose** due to author differences.

Pungent: Terminating gradually in a hard sharp point. (pungens)

Pubescent: Generally hairy; more specifically covered with very short, weak, dense hairs.

Puberulent: Minutely pubescent. Covered with minute soft hairs, almost imperceptible, at right angles to the surface.

Rhombic: Oval, a little angular in the middle.

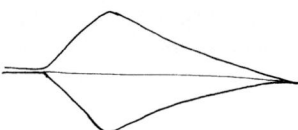

Rugose: Wrinkled. Covered with reticulated lines, the spaces between which are concave.
Rugulose: Finely rugose (wrinkled by irregular lines and veins)
Serrate: Having sharp (more or less) straight-edged teeth pointing towards the apex. See also **Dentate** and **Denticulate**.

Serrulate: Minutely serrate.
Setaceous: Either bearing bristles or bristle like.
Scabrous: Slightly covered with short hardish points; rough to the touch.
Sessile: Stalkless
Spathulate: spatula shaped. A bit like *Bellis perennis* leaves.
Stellate: Hair structure composed of 5 or more arms. See Appendix 4.
Sub... : Not completely, somewhat.
Subulate: Awl shaped. Tapering from a narrow or moderately broad base to a very fine point.
Tomentose: Heavily felted or woolly. In *Phlomis* tomentose is usually stellate-tomentose. The difference between tomentose and **Lanate** (**Lannose**) or **Pannose** is (in descriptions of *Phlomis) small* and usually author dependant Tomentose should be reserved for the densest covering.
Truncate: Terminating very abruptly, as if a piece had been cut off.

Villous: Having a shaggy appearance due to a covering of long soft hairs.
Viscid: Sticky to the touch.

Appendix 4—*Phlomis* Hair Types

Hair types found in *Phlomis* species. They vary from simple, multi-jointed, stellate, stalked stellate, stellate and stalked stellate with long hair, dendroid or tree-like, to glandular versions of these (bottom row).

Appendix 5—References

Adylov T. A., **Kamelin**, R. V. & **Makhmedov**, A. M. (1987) *Opred. Rast. Sred. Azii* 82-113

Bentham, G. (1834) *Labiatarum Genera et Species*: 323-644

Boissier, P. E. (1879) *Phlomis* in *Flora Orientalis*, 4:779-793

Brantjes, N. B. M. (1981) *Floral Mechanics in Phlomis (Lamiaceae)* in *Ann. Bot. London* 47:279-282

Briquet, J. (1895-7) In A. Engler & K. Prantl, *Die natürlichen Pflanzenfamilien*, 4: 414-416

Compton J. (1987) *Success with Unusual Plants*. Collins

Huber-Morath, A. (1958) *Die anatolischen Arten der Gattung Phlomis Linnaeus* in *Bauhinia* 1(2): 97-123

Kamelin, R. V. & **Makhmedov** A. M. (1990) *A New System of the Genus Phlomis (Lamiaceae)* in *Bot. Anicheskii Zhurn. AL* (Moskva) 75:1163-1167

Komarov (1954) Flora of the URSS (English trans. by O. E. Knorring) 21:42-77

Link, J. H. F. (1829) *Handbuch zür Erkennung der nutzbarsten und am häufigsten vorkommenden Gerwächse*, 1:479

Ludwig, W. (1968) *Phlomis tuberosa, P. russeliana und P. samia* in *Hessische Floristische Briefe* 17,196:19-22

Mateu, I. (1986) *Revision del Genero Phlomis Linnaeus (Labiatae) en La Peninsula Iberica e Islas Baleares* in *Acta Botánica Malacitana*, 11:177-204

Moench, C. (1794) *Methodus plantas Horti botanici et agri Margurgensis a straminium situ describendi.*

Post, G. E. (1933) *Flora of Syria, Palestine and Sinai*, ed 2,2: 323-389

Rivas Goday S & **Rivas Martínez** S. 1969. Matorrales y tomillares de la Península Ibérica comprendidos en la clase Ononido-Rosmarinetea Br.-Bl. *Anal. Inst. Bot. Cavanilles*, 25: 5-201, Madrid

Vierhapper, F. (1915) *Beiträge zur Kenntnis der Flora Kretas* in *Österreichische botanische Zeitschrift* 89: 257-299

Indexes

If you know which country the *Phlomis* comes from, then look on pages 70–79 and you will find that each species described in this book is marked in bold type and has the page reference on the far left.

If the plant is a shrub, then these are arranged alphabetically from page 16–36.

If the plant is a perennial, these are split into two sections.
If the corolla is that of section *Phlomis* (see diagram on page 7), these are listed alphabetically from pages 37–59.
If the corolla corresponds to that of section *Phlomoides* (see diagram on page 7), these are listed alphabetically from 60–69.
See also 'A Guide to identifying *Phlomis* using this book' on page 14

The Index by country starting on page 70, has the Authors names given in full.

Key:

✦ means that the plant is in the NCCPG National Collection.

✧ means that the plant is in the French National Collection.

❀ means a good garden plant in the author's opinion.

❀❀ means an outstanding garden plant in the author's opinion.

❄ means that the plant may be killed with sub-zero (°C) temperatures.